Forensic DNA Analyses Made Simple

Sequencing genetic material is now common practice. The general population have become consumers of this information but without an understanding of the biological processes that render sequencing data useful. The interpretation of genetic sequence depends on an appreciation of the basics of genetics and the limits of such data. This book provides the background necessary to understand, interpret, and apply sequencing information to real-world problems. Replication of genetic material, the structure of DNA, typing methods, and forensic applications are all discussed in this useful primer.

Key Features

- Provides self-learning about DNA fingerprinting.
- Includes sections on how to analyze and interpret DNA fingerprinting.
- Covers legal and medicolegal issues and case analyses.
- Teaches novice legal community about DNA fingerprints.
- Summarizes for a general audience the role of ancestry, DNA, and what that means.

Forensic DNA Analyses Made Simple
A Guide for the Curious

Omar Bagasra and Ewen McLean
Illustrated by Mohammad Saffar

CRC Press
Taylor & Francis Group
Boca Raton London New York

CRC Press is an imprint of the
Taylor & Francis Group, an **informa** business

Designed cover image: Mohammad Saffar

First edition published 2024
by CRC Press
6000 Broken Sound Parkway NW, Suite 300, Boca Raton, FL 33487–2742

and by CRC Press
4 Park Square, Milton Park, Abingdon, Oxon, OX14 4RN

CRC Press is an imprint of Taylor & Francis Group, LLC

© 2024 Omar Bagasra, Ewen McLean, and Mohammad Saffar

Library of Congress Cataloging-in-Publication Data
Names: Bagasra, Omar, 1948– author. | McLean, Ewen, author.
Title: Forensic DNA analyses made simple : a guide for the curious / Omar Bagasra and
    Ewen McLean; illustrated by Mohammad Saffar.
Description: First edition. | Boca Raton, FL : CRC Press, 2023. | Includes bibliographical
    references and index.
Identifiers: LCCN 2022056277 (print) | LCCN 2022056278 (ebook) | ISBN 9781032022345
    (hbk) | ISBN 9781032015040 (pbk) | ISBN 9781003182498 (ebk)
Subjects: LCSH: DNA fingerprinting. | Forensic genetics—Technique. | DNA—Analysis.
Classification: LCC RA1057.55 .B34 2023 (print) | LCC RA1057.55 (ebook) |
    DDC 614/.1—dc23/eng/20230420
LC record available at https://lccn.loc.gov/2022056277
LC ebook record available at https://lccn.loc.gov/2022056278

ISBN: 978-1-032-02234-5 (hbk)
ISBN: 978-1-032-01504-0 (pbk)
ISBN: 978-1-003-18249-8 (ebk)

DOI: 10.1201/9781003182498

Typeset in ITC Garamond Std
by Apex CoVantage, LLC

Contents

Preface

Using DNA to solve crimes and find one's ancestry has become one of the most common practices of our age. Anytime we watch crime dramas on TV, we see that law enforcement and detectives utilize DNA fingerprints to catch or exonerate someone accused of committing a crime. Similarly, there are many TV shows where individuals find their ancestors by utilizing DNA analytical services. However, not many people really know what DNA analysis is and what it entails. How is the DNA fingerprinting process carried out, and how do we find out where our forefathers came from? Surprisingly, not even law enforcement and defense attorneys who represent individuals accused of committing crimes understand much about DNA, let alone its analyses. What does it take to figure out how DNA is analyzed, what are the methods used, and where can things go wrong? In this book, we explain the process of DNA analysis in a simplified manner. The text has been designed so that anyone with a fifth-grade education can learn the complete DNA-based identification process. The book starts with the understanding of a single cell. All life is based on cells. As humans, we have 37.2 trillion cells—more than 100 billion neurons are in our brains. After explaining a single cell, we explain how our cells divide; this is one of the most critical parts of the book, because a thorough understanding of cell division, mitosis, and meiosis is essential to understand ancestry DNA as well as solve crimes such as gang rape. This critical chapter is followed by numerous questions to ensure that the reader grasps the process. This self-learning module is designed so that as a reader you can test yourself before moving forward.

The next chapter covers the structure of DNA and why and how it is used to solve crimes. This is followed by a clear explanation of short tandem repeats (STR), a critical chapter to discover, and how DNA fingerprinting is carried out. This is followed by a section that explains how the gender of a DNA sample is determined. Then, we explain the CODIS database that the FBI uses.

Once these introductory chapters are completed, you will learn how DNA is extracted and typed. You will also appreciate how DNA is collected from a crime scene and protected to avoid any contamination and how DNA evidence is processed. After this, you will learn about how statistical analyses are used in the courtroom. DNA of all *Homo sapiens* is extremely similar but varies at repeated sequences or what we call short tandem repeats, or STRs. Only genetic twins or triplets have the same STRs. Therefore, matching the STRs is a task that requires some mathematics, most importantly understanding of probability and statistics. A DNA match may not have any meaning if the case is not presented in the form of probability and statistical analyses. To fully grasp the concept, we have explained how DNA is used in solving crimes by case analyses. Finally, we explain autosomal dominant versus autosomal recessive and X-linked disorders.

We are hopeful that this book, full of illustrations, will be of value to everyone as well as a valuable resource for defense attorneys and other law enforcement individuals.

Author Biographies

 **Omar Bagasra, MD, PhD** In 1948, somewhere on the plains of India, Omar Bagasra was born in the back of a wooden oxcart. His refugee family was migrating north during the exodus of the 25 million souls who were forced to leave their ancestral homelands when the former British colony of India was being partitioned during its struggle to become independent. At least 8 million of these refugees—Sikhs, Muslims, and Hindus alike—perished in this partitioning. Being Muslims, Omar's family settled in the new nation of Pakistan, where his father became a successful grain merchant and where ten more brothers and sisters were born and one was adopted. In this somewhat-volatile environment, Omar grew into a young man.

At the age of 16, Omar decided to study other faiths and adopted an ascetic lifestyle—starting as a Buddhist monk, a creed for which warfare is anathema. He left his parents' home in Pakistan and journeyed to a monastery in Tibet where he lived as a Buddhist disciple and then moved to northern provinces and Afghanistan and visited many *faqirs*. After two years of a life of a *faqir*, Omar reflected that the *scientific* understanding of nature was just as important as the path to truth as the more mystical consciousness approach of the ascetic monks.

He therefore returned to Pakistan and enrolled in the Karachi University, where he earned a bachelor's and a master's degree in biochemistry. "I wanted to get even higher education," he says, "but in Pakistan at that time, that was as high as I could get." So in 1972, he flew to Chicago's O'Hare airport, carrying just a suitcase of clothing and an extra hundred dollars in his pocket.

Omar didn't know anyone in the United States, but he soon found employment in the road construction industry, and he learned to speak better English—his seventh language. He then got better job manufacturing brake shoes for the Ford Motor supplier in Albion, Indiana, near Ft. Wayne. Omar saved his wages and enrolled first at Indiana University and, after a year, at the University of Louisville; soon he got his first scientific job working as a lab technician at the nearby Clark County Memorial Hospital in Jeffersonville, Indiana. There he met a young nurse, Theresa Mahoney, and the two were married. By 1980, Omar had earned a PhD in microbiology and immunology. He joined a group in Albany, New York, to do his postdoctoral fellowship in infectious disease, and the family moved to Philadelphia when his postdoc mentor moved to the city, where Omar became a junior faculty member at Hahnemann University and a citizen of the United States of America.

Soon thereafter, Dr. Bagasra decided to go to medical school. But admissions policies at that time were very restrictive for individuals born and educated overseas and the tuition was more than he can afford, so the 32-year-old Omar—never one to be confined by national borders—went to study medicine at the Universidad Autónoma in Ciudad Juarez, Mexico. After two years of study, he went to Temple University, where he completed his clinical training. Subsequently, he completed residency in anatomic pathology at Hahnemann and Temple Universities, a fellowship in clinical laboratory immunology at the Saint Christopher's Hospital for Children, while serving as a full-time faculty member at Hahnemann University, all in Philadelphia.

Before coming to Claflin University, Dr. Bagasra held professorships at Hahnemann University (1980–1987) and Thomas Jefferson University in Philadelphia, where he served as Director of the Molecular Retrovirology Laboratories and Section Chief of Molecular Diagnostics of the Center for the Study of Human Viruses, as well as Professor of Medicine from 1991 to 1998. Dr. Bagasra also keeps a hand in clinical work—he is currently board-eligible in anatomic pathology and a diplomat of the American Board of Medical Laboratory Immunology (ABMLI) and the American Board of Forensic Examiners and a fellow of the American College of Forensic Examiners.

Dr. Bagasra's research interests have long been associated with the study of HIV and AIDS. In fact, he has been on the trail of the virus since 1981—the year of the first scientific report. For several years, he focused on trying to gain insight into the molecular pathogenesis of HIV and the role of microRNA in the protection against lentiviruses. In 1998, he was the first to clearly discuss the protective role of small RNAs against retrovirus and lentivirus ("HIV and Molecular Immunity"). His unswerving dedication to his work has resulted in over 200 scientific articles, book chapters, and books. During the last few years, he has received several prestigious national and international awards and recognitions. In 2002 and in 2014, he received the Faculty Scholar Award from the American Association for Cancer Research. In 2006 he was the corecipient of the South Carolina Governor's Award for Excellence in Science. From 2002 to 2006, he also served as Council Member of the American Association of Cancer Research (MICR-AACR). Dr. Bagasra currently serves as Professor of Biology and Director of the South Carolina Center for Biotechnology at Claflin University, which he founded in 2001. Much of his work has been recognized in top-tier journals, such as the *New England Journal of Medicine, The Proceedings of the National Academy of Science, Journal of Virology, Journal of Immunology,*

Journal of Pediatrics, Nature Medicine, Nature Protocol, Science, EBioMedicine, among others.

The Institut Pasteur's Luc Montagnier—the discoverer of the AIDS virus and a 2008 Nobel Laureate—described Dr. Bagasra as "a skilful researcher . . . [and] a discerning scholar who explores new ideas," observing he already had a track record for challenging conventional wisdom and being proved correct. "Every scientist now knows that a significant percentage of circulating lymphocytes are infected with HIV but in 1992 his findings were highly controversial."

Currently, Dr. Bagasra has been working on the etiologies of autism as well as Alzheimer's disease. He has discovered several neurogenic factors from plants that may remedy neurodegeneration and dementia. He recently published a book on the role of environmental chemicals in the pathogenesis of autism.

 **Ewen McLean, PhD** Originally from Peterborough, UK, Dr. McLean completed his doctorate in biomedical sciences at the University of Bradford. He has served as Research Leader in biotechnology at Aalborg University, Denmark, and has held positions as Department Chair, Program Professor, and Director at various institutes in the United States, Middle East, and Caribbean. He has actively

participated on national research review panels and contributed to various committees and commissions on a global basis. He has mentored 40 graduate students and has served on the editorial boards of nine peer-reviewed journals. He has published over 110 peer-reviewed papers and 220 other contributions. Since 2013, he has been based in Columbia, South Carolina.

Mohammad Saffar, Illustrator Mohammad Saffar is a graphic designer with ten years of experience in graphic design and animation. He earned a BA in graphic design at Soore University of Tehran/Iran. After working for ten years in graphic design and animation in the industry for companies in India, Oman, and Iran, he went to the United States. He received his master of fine arts degree in digital production art. Mohammad likes to learn new technology and share that with students. Accordingly, he spends time teaching graphics and animation at universities in person and online.

Introduction

Sometimes it's difficult to appreciate the importance of basic science material to forensic science. If you are a legal, a paralegal, or just curious to understand and interpret DNA fingerprint analyses or want to learn how your ancestry was carried from the DNA sample you sent to one of many ancestry databases, we are sure you've felt the frustration in understanding your results. All the relevant concepts of DNA science are based on simple biological facts. Therefore, by understanding some simple biological concepts, especially meiosis, you will be able to get a clearer idea of how genes are inherited. Some folks may think that if one knows that the egg and the sperm each end up with half the usual number of chromosomes, it is sufficient.

Today we know much more about human inheritance than we knew 50 years ago, and our knowledge is increasing rapidly. You may hear about many aspects of inheritance, ranging from normal variations among your children (height, baldness, intelligence, skin color, etc.)

DOI: 10.1201/9781003182498-1

to the chances of having, say, a child with Down's syndrome. In fact, all these problems are intimately involved with the minute events of meiosis and can, incidentally, be easily understood if you take the few minutes necessary to master the basic principles. If you are in a legal or law enforcement area, you may be the only available source of counseling to your bewildered clients and, in most straightforward cases, should be able to provide reasonably intelligent guidance if armed with a sound comprehension of simple genetic processes such as meiosis. It's that simple, and it provides a good example of how basic science can be truly relevant to helping people make intelligent decisions. It's our purpose to provide you with the basic facts about a few genetic processes that will help you understand how DNA fingerprinting and inheritance work and how we inherit genetic markers and how DNA fingerprints are interpreted.

There is one fundamental genetic mechanism, called **meiosis**, that you need to master. Once you have a grasp of this process, you will be able to apply it to human genetics and understand how it can be used in solving crimes and ancestry links. It's worth emphasizing that if you understand meiosis, there is very little about the human inheritance that you won't be able to comprehend.

Chapter 1

The Cell

All living things on the planet are made of cells. Each cell can be visualized as being like a brick that makes a building. Most of the living things on Earth are single-cell life-forms. Bacteria are the most common single-cell life-forms on Earth. There are innumerable single-cell life-forms in the ocean and in rivers and lakes. All cells share the same demands: the need to get energy from their environment, respond to their surroundings, excrete waste products, and reproduce. Cells must also be able to separate their relatively stable interior environment from the ever-changing external environment. They do this by coordinating many processes that are carried out in different parts of the cell. Structures that are common to many different cells indicate the common history shared by cell-based life. There are two major groups of life-forms: prokaryotes and eukaryotes. Bacteria are prokaryotes and have no nuclear organelle that contains DNA (short for *deoxyribonucleic acid*), whereas a eukaryote's DNA structures are surrounded

DOI: 10.1201/9781003182498-2

by a nuclear membrane that isolates their DNA from the rest of the internal environment of the cytoplasm. A single cell is like a balloon filled with liquid (cytoplasm) and multiple small balloon-like structures (organelles), as well as numerous stringlike shapes, each of them serving specialized functions (**Figure 1.1**). There are two structures that contain DNA: the nucleus and the mitochondria. The nucleus of cells houses genomic DNA, and it codes all the functions of a cell. Humans have a total of 46 chromosomes located inside the nucleus, and these threadlike structures carry genetic information in the form of genes. Generally, an intact nucleus is needed to identify one human from another and from other

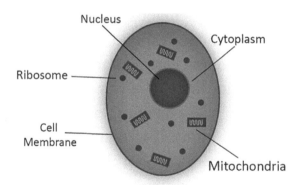

Figure 1.1 Without the use of special dyes, we cannot make out the microstructure of cells by looking at them through a normal microscope. It is only with an electron microscope that the true microstructure of a cell becomes apparent. A typical human or animal cell will look like the preceding diagram and contain the following components.

life-forms on Earth. Genes regulate all the cell's biological functions. Every species has unique DNA sequences (DNA codes); thus, the DNA sequences of humans differ from that of chimpanzees, and every species on Earth carries its own unique DNA sequences. Humans each have many unique DNA signatures or codes that are used to differentiate between two individuals. Only identical twins (maternal twins) have identical DNA fingerprints. Every cell in the human body contains DNA, so anything that contains cells, including saliva, semen, urine, feces, blood, air, and so on, also contains a certain amount of intact human DNA, no matter how small, that can be used for DNA analyses. A forensic expert must learn how to collect DNA from a crime scene and how to identify and interpret the results from samples. The same DNA analytical method is used to identify the biological father or mother. This method is known as STR, or short tandem repeat, analysis.

- *Nucleus.* Almost all cells have a nucleus. A nucleus contains the genetic information for cells. A human nucleus will contain the genetic information in 46 chromosomes (we will be looking at chromosomes, DNA, and genes in more detail later). Genes control the activities that occur in the cell, such as what proteins or enzymes to make.
- *Cytoplasm.* A gel-like substance that makes up most of the cell. It is made up of many different substances.

- ■ *Cell membrane.* The cell membrane is the outside of the cell. It provides a boundary between the cytoplasm and the external environment. The cell membrane decides which substances come in and leave a cell. It is partially permeable (meaning, that some substances can pass through it).
- ■ *Mitochondria.* These are found in the cytoplasm of all living cells. They are responsible for some of the chemical reactions that take place in respiration, which releases energy that the cell can use. Cells that require a lot of energy have a lot of mitochondria.
- ■ *Endoplasmic reticulum.* In addition to the cell membrane, there is a network of membranes in the cytoplasm called the endoplasmic reticulum.
- ■ *Ribosomes.* The endoplasmic reticulum is covered by ribosomes. Ribosomes are responsible for assembling proteins for the cell.

Cell division, cell differentiation, and cell death are the three key processes that control the state of balance between body systems for multicellular organisms to survive and function properly. The growth and survival of cells, and the integrity of the genome, are ordered by a complex network of pathways in which cell cycle checkpoints, DNA repair, and programmed cell death have a key role. Disruption of genomic integrity and compromised regulation of cell death may both lead to uncontrolled cell growth. Compromised cell death can also favor genomic instability. It is becoming clearer that dysregulation of the cell cycle and cell death processes play an important role in the development of major disorders, such as cancer, cardiovascular disease, infection, inflammation, and neurodegenerative diseases, such as dementia. Research in these areas has led to the design of new approaches for the treatment of various conditions associated with abnormalities in the regulation of cell cycle progression or cell death. Every organism on Earth, including plants and animals, must reproduce and grow. All multicellular organisms grow from single-cell embryos to their respective normal size. Except in some special circumstances, the larger the animal, the more cells they have. The human body is made of trillions of cells. Most of the time, an organism grows by cell multiplication using a process called mitosis. This is an asexual cell division. In highly specialized organs, animals produce cells that participate in sexual reproduction. In humans, the male's testicles and the female's ovaries produce gametes (sperms and eggs) by a mechanism called meiosis.

Humans acquire their mitochondrial DNA **ONLY** from their mother, whereas the Y chromosome of a male child **ONLY** comes from a biological father, and both are used to develop a family tree and ancestry identification.

Therefore, mitochondrial analysis traces a person's maternal lineage: mother, maternal grandmother, maternal great-grandmother, and so on. Similarly, Y chromosome testing traces only one line of a person's male ancestry, starting with a man's father, his paternal grandfather, his paternal great-grandfather, and so forth.

Lineage testing can trace your ancestry back to individuals who carried a particular DNA type throughout prehistory until the present day.

To attain a level of proficiency in DNA analyses, you must first understand the meiotic process and how it is related to human inheritance. Accordingly, we must define a few subobjectives which will lead you to handle the overall objectives. Don't be afraid; put your learning cap on and start to absorb the fundamental knowledge of inheritance. In the next chapter, you will master the following concepts.

1. The basic differences between meiosis and mitosis.
2. The genetic consequences of the meiotic process.
3. The difference between segregation and independent assortment.
4. How meiosis differs between males and females.
5. The difference between reductional and equational division, and where in the body these processes occur.
6. When crossing over occurs and in which cells.
7. The genetic consequences of crossing over.
8. Under what conditions, if any, reduction division can occur at meiosis II.

9. The components of the synaptonemal complex and where they develop.
10. The significance of the synaptonemal complex.

The preceding objectives and terms may sound scary, but most middle- and high-schoolers know all about this!

For further information refer to Bibliography 1–4.

Chapter 2

How Cells Divide and Why It Is Important to Know the Process

2.1 Self-Learning Cycle #1: Chromosomes and Independent Segregation

Our genes control the activities of all cells by directing protein synthesis. Genes are carried by chromosomes, which are distributed to one's offspring through the meiotic process. Your child receives half his/her complement of chromosomes from you, and the other half from your mate. The half he or she receives from you is only half of *your* whole set (the same for your mate) and is, to some extent, a chance of happening. There are two main components to meiosis, one *fixed* and one *random*. Here's how it works: **chromosomes come in pairs**.

In each of your cells (except mature germ cells), there are 23 pairs—46 chromosomes—consisting of one pair of *sex* (X and Y determining) *chromosomes* and 22 pairs of *autosomal chromosomes* (**Figure 2.1**). One chromosome of each pair of *homologous chromosomes* was contributed to you from your mother (maternal chromosome), and the other from your father (paternal chromosome). These half (haploid) sets were derived from the egg of your mother and the sperm of your father. You have also received another type of DNA, called mitochondrial

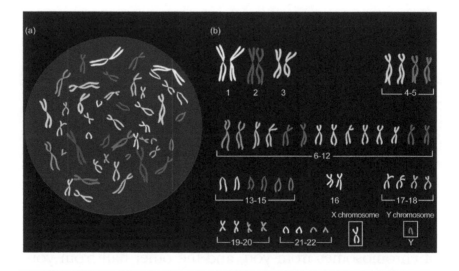

Figure 2.1   Humans have 46 chromosomes, and (a) shows each of them in a cell just before it divides into two. Each "X-shaped" figure represents a single pair of chromosomes. The chromosome images can be cut and then arranged in groups according to kinetochore position (a kinetochore is a disc-shaped protein structure associated with duplicated chromatids in eukaryotic cells, where the spindle fibers attach during cell division to pull sister chromatids apart).

DNA, that is only contributed by your mother (we will get into this in detail later). Therefore, all 23 pairs of (46) chromosomes were created through the meiotic process that occurred in your father's testis and your mother's ovary; at that time, the first seed of your being came into existence (known as fertilization).

The *fixed* component of meiosis means that during the formation of the egg or sperm (the *gametes*), each gamete will receive only one member of each homologous pair of chromosomes, never both. In other words, the maternal and paternal chromosomes of each pair separate into different gametes. This is called *segregation* (**Figure 2.2**).

Many of our inherited characteristics are determined by the interaction of a single pair of genes that occupy identical places (*loci*) on a pair of homologous chromosomes. The two genes may be alike or different, and they may interact in different ways. Since they are integral components of chromosomes, they necessarily accompany the chromosomes into the gametes during meiosis. The offspring formed by the union of a male and a female gamete will then receive one of the two genes of a particular *locus* from the father's *genome* (set of genes) and one of the mother's two genes occupying the same locus. Just which one of the mother's two genes, and which one of the fathers, is a chance phenomenon.

Note regarding chromosome numbers: A bacteria usually contains a single small circular chromosome. But eukaryotic cells have many large linear chromosomes. The

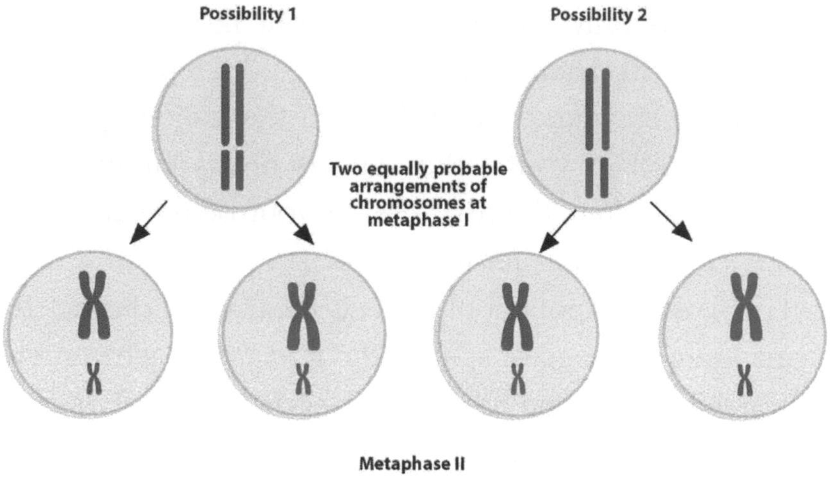

Figure 2.2   Segregation. This figure summarizes *segregation*, which means only one chromosome of each pair. It is a matter of chance (*independent assortment*) whether it is the maternal or paternal chromosome. Since both processes occur simultaneously, we sometimes lump them together and call the whole thing "independent segregation." But you should bear in mind that *independent segregation* has two components, *segregation* and *independent assortment*. The former ensures that the gametes will have only half the number of chromosomes of the organism and, in particular, one representative of each homologous pair. This means that the offspring will have copies of all the genes necessary to maintain the species' characteristics. *Independent assortment* provides for *genetic variation* within that species' framework by making it highly unlikely that any two gametes from a single person will have the same combination of maternal and paternal chromosomes. Actually, there is only one chance in 2^{23} (one in more than 8,000,000) that any two sperm or eggs will be alike. Since the same chance exists for the other gamete and the fertilization of a particular egg by

Figure 2.2 (Continued) **a particular sperm is also a chance phenomenon, the chance that any two brothers and sisters will be exactly alike is one in 4^{23} (one in >70 billion); so you can see that it is an independent assortment that keeps us all from being monotonously alike.**

number and DNA content vary significantly between species. As a rule, the more complex the organism, the bigger the size of the genome. However, there are exceptions. For example, onion and lily contain, respectively, about 5 and 30 times as much DNA as that of a typical human cell. Similarly, the chromosome numbers vary among organisms and do not correlate with the evolutionary tree.

2.1.1 Mitosis vs. Meiosis

Organisms grow and reproduce through cell division. In eukaryotic *cells*, the production of new cells occurs because of *mitosis* and *meiosis* (**Figure 2.3**). These two nuclear division processes are similar but distinct. Both processes involve the division of a *diploid cell*, or a cell containing two sets of *chromosomes* (one chromosome donated from each parent).

In **mitosis**, the genetic material (*DNA*) in a *cell* is duplicated and divided equally between two cells. The dividing cell goes through an ordered series of events, called the *cell cycle*. The mitotic cell cycle commences in the presence of growth factors or other indicators that

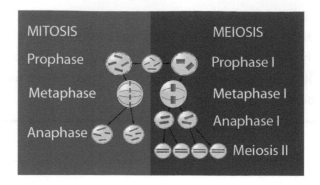

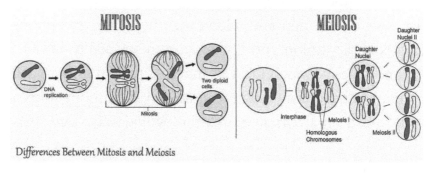

Differences Between Mitosis and Meiosis

Figure 2.3 **This figure illustrates the major differences between mitosis and meiosis.**

signal that new cells need to be produced. Somatic *cells of the body* replicate by mitosis. Examples of somatic cells include *fat cells*, *blood cells*, skin cells, or any body cell that is not a *sex cell*. Mitosis is necessary to replace dead cells, damaged cells, or cells that have short life spans.

1. Cell Division

– **Mitosis:** A somatic cell divides **once**. Cytokinesis (the division of the cytoplasm) occurs at the end of telophase.

- **Meiosis:** A reproductive cell divides **twice**. Cytokinesis happens at the end of telophase I and telophase II.

2. Daughter Cell Number

- **Mitosis: Two** daughter cells are produced. Each cell is diploid, containing the same number of chromosomes.
- **Meiosis: Four** daughter cells are produced. Each cell is haploid, containing one-half the number of chromosomes as the original cell.

3. Genetic Composition

- **Mitosis:** The resulting daughter cells in mitosis are genetic clones (they are genetically identical). **No recombination or crossing over occurs**.
- **Meiosis:** The resulting daughter cells contain different combinations of genes. **Genetic recombination occurs** because of the random segregation of homologous chromosomes into different cells and by the process of crossing over (transfer of genes between homologous chromosomes).

4. Length of Prophase

- **Mitosis:** During the first mitotic stage, known as prophase, chromatin condenses into discrete chromosomes, the nuclear envelope breaks down, and spindle fibers form at opposite poles of the cell. A cell spends less time in prophase of mitosis than a cell in prophase I of meiosis.
- **Meiosis:** Prophase I consist of five stages and lasts longer than the prophase of mitosis. The five stages

of meiotic prophase I are leptotene, zygotene, pachytene, diplotene, and diakinesis. These five stages do not occur in mitosis. Genetic recombination and crossing over take place during prophase I.

5. Tetrad Formation

2.1.2 *Differences in Mitosis and Meiosis*

While the processes of mitosis and meiosis differ in some ways, they also have similarities: both processes have a growth period, called **interphase**, in which a cell replicates its genetic material and organelles in preparation for division (**Figure 2.4**).

Both mitosis and meiosis involve phases: **prophase, metaphase, anaphase**, and **telophase**. During meiosis, a cell goes through these phases twice. Each process also involves the lining up of duplicated chromosomes, known as sister chromatids, along the metaphase plate, which is an imaginary line equidistant from the point at which the centromeres of chromosomes pair at the spindle equator. This happens in the metaphase of mitosis and metaphase II of meiosis.

In addition, both mitosis and meiosis require the separation of sister chromatids to form daughter chromosomes. This event occurs in the anaphase of mitosis and anaphase II of meiosis. Finally, both processes end with the division of the cytoplasm to produce individual cells.

Sexual reproduction requires that each parent pass only half of their genetic material to their offspring. To do this, the parent had to evolve a new method of creating a

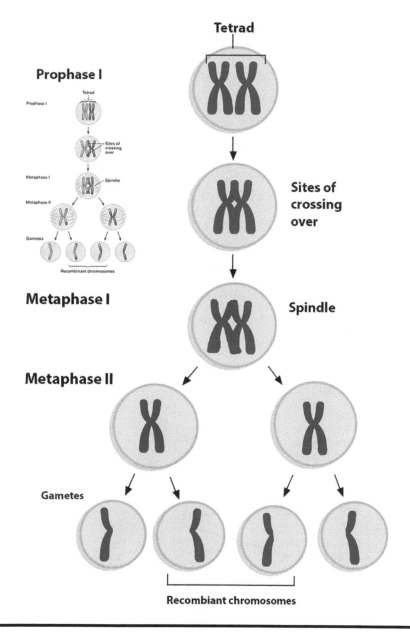

Figure 2.4   Various stages of meiosis.

package (that we call a **gamete**) containing precisely half of their genetic material (chromosomes).

Mitosis and **meiosis** are nuclear division processes that occur during cell division.

Mitosis involves the division of body cells, while meiosis involves the division of sex cells.

Two daughter cells are produced after mitosis and cytoplasmic division, while **four daughter cells** are produced after meiosis.

The division of a cell occurs once in mitosis, but twice in meiosis.

Daughter cells resulting from mitosis are **diploid**, while those resulting from meiosis are **haploid.**

Daughter cells that are the product of mitosis are genetically identical. Daughter cells produced after meiosis are genetically diverse.

Tetrad formation occurs in meiosis, but not mitosis (**Figure 2.5**).

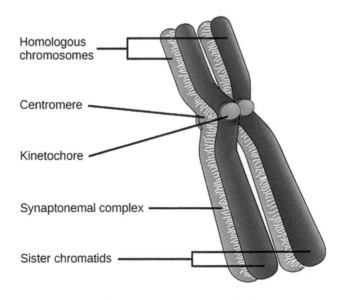

Figure 2.5 Tetrad formation.

Summary of Differences between Mitosis and Meiosis

Mitosis	Meiosis
Equal division (i.e., diploid chromosomes).	Reduction division (haploid chromosomes).
Purpose is to multiply cells.	No multiplication of cells.
Asexual division, somatic cell division.	Sexual division occurs in sex cells only.
Involves one cell division with 4 stages, that is, *prophase*, *metaphase*, *anaphase*, and *telophase*.	Involves two successive divisions that result in the reduction of chromosomes numbers to 50%.
Results in two daughter cells, each with pair of chromosomes.	Results in four cells, with each cell with 50% less number of chromosomes.
During metaphase, the chromosomes align at the equatorial plate of the cell. This event is due to the presence of kinetochore microtubules that pull these chromosomes to the opposite sides.	During metaphase I, homologous chromosomes begin to align themselves at the equatorial plate as they bind to the mitotic spindle. During metaphase I, single chromosomes align at the equatorial plate after each cell has completed forming the spindle fibers.
During the anaphase stage, each single-stranded chromosome pair is segregated toward the opposite poles of the cell. This activity is initiated by the mitotic spindle.	During anaphase I, double-stranded chromosomes are separated toward each cellular pole. During anaphase II, the segregation of chromosomes takes place.

During the telophase stage, the complete transfer of genetic material from the parent cell to the two daughter cells takes place and new nuclear membranes are formed.	At the end of telophase I, each daughter cell carries a haploid set of chromosomes. Telophase II is similar to the telophase in mitosis. Half of the genetic materials is transferred to the four daughter cells.
No genetic variations in the two daughter cells occur.	In meiosis, segregation and independent assortment and gene crossover contribute to genetic variations in each of the four daughter cells.

2.1.3 Mitosis and Meiosis Similarities

While the processes of mitosis and meiosis express several differences, they are also similar in many ways. Both have a growth period, called **interphase—the point at which the** cell duplicates its genetic material and organelles in preparation for division.

Both mitosis and meiosis involve phases: **prophase**, **metaphase**, **anaphase**, and **telophase**; but in meiosis, the cell cycle phase is undertaken twice. Both processes also include the lining up of individual duplicated chromosomes, known as sister chromatids.

The *random* component says simply that the segregation pattern of each pair of chromosomes is independent of every other pair. This is an *independent assortment*. It means that if the maternal chromosome of pair A ends up in a particular sperm, it may, by chance, end up with either the maternal *or* the paternal chromosome of pair B, and so forth, for every chromosome pair.

To summarize, *segregation* means only one chromosome of each pair. It is a matter of chance (*independent assortment*) whether it is the maternal or paternal chromosome (**Figure 2.6**).

Since both processes occur simultaneously, we sometimes lump them together and call the whole thing "independent segregation." But you should bear in mind that

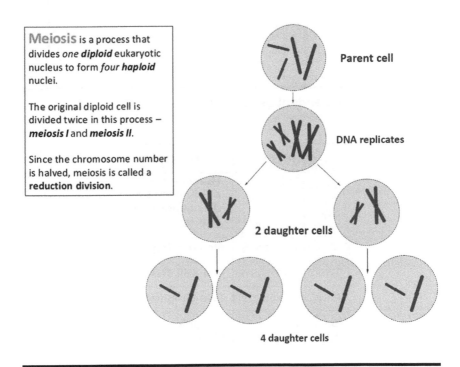

Meiosis is a process that divides *one diploid* eukaryotic nucleus to form *four haploid* nuclei.

The original diploid cell is divided twice in this process – *meiosis I* and *meiosis II*.

Since the chromosome number is halved, meiosis is called a **reduction division**.

Parent cell

DNA replicates

2 daughter cells

4 daughter cells

Figure 2.6   Summary of meiosis.

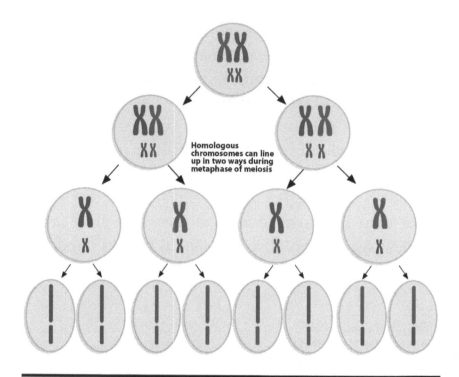

Figure 2.7   Independent assortment.

MEIOSIS	MITOSIS
Homologous chromosomes pair up	Homologous chromosomes do not normally pair up
Crossing over	No crossing over
Two cell divisions	One cell division
Four daughter cells	Two daughter cells
Daughter cells haploid (n)	Daughter cells diploid (2n)

Figure 2.8   Differences between mitosis and meiosis.

independent segregation has two components, *segregation* and *independent assortment* (**Figures 2.7** through **2.9**). The former ensures that the gametes will have only half the number of chromosomes of the organism and, in particular, one representative of each homologous pair. This means that the offspring will have copies of all the genes necessary to maintain the species' characteristics. *Independent assortment* provides for *genetic variation* within that species' framework by making it highly unlikely that any two gametes from a single person will have the same combination of maternal and paternal chromosomes. There is only one chance in 2^{23} (one in more than 8,000,000) that any two sperm (or eggs) will be alike. Since the same chance exists for the other gamete

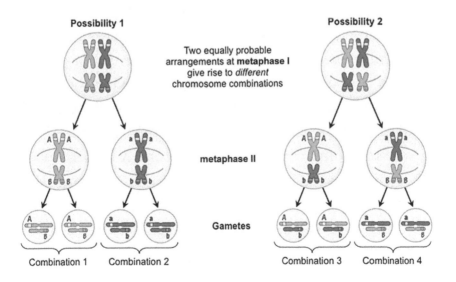

Figure 2.9 Segregation and independent assortment.

and the fertilization of a particular egg by a particular sperm is also a chance phenomenon, the chance that any two brothers and sisters will be exactly alike is one in 4^{23} (one in 700 billion that any two sperm or egg will be alike); so you can see that it is independent assortment that keeps us all from being monotonously alike. Of course, there are additional mechanisms that create even more differences between each of the germ cells.

Many of our inherited characteristics are determined by the interaction of a single pair of genes which occupy identical places (*loci*) on a pair of homologous chromosomes. The two genes may be alike or different, and they may interact in different ways. Since they are integral components of chromosomes, they necessarily accompany the chromosomes into the gametes during meiosis. The offspring formed by the union of a male and a female gamete will then receive one of the two genes of a particular *locus* from the father's *genome* (set of genes) and one of the mother's two genes occupying the same locus. Just which one of the mother's two genes, and which one of the father's, is a chance phenomenon.

Note regarding chromosome numbers: A bacteria usually contains a single small circular chromosome. But eukaryotic cells have many large linear chromosomes. The number and DNA content vary significantly between species. As a rule, the more complex the organism, the bigger the size of the genome. However, there are exceptions. For example, the onion and lily contain about 5 and 30 times more DNA, respectively, than that of a typical human cell. Similarly, chromosome numbers vary considerably among organisms and do not correlate with the evolutionary tree. For example, orchids have over 100 chromosomes.

2.1.4 Ancestry Tracing

Biogeographical ancestry can be traced using a technique called admixture testing. This test focuses on the 22 pairs of nonsex (autosomal) chromosomes in each cell. Since the cell's chromosomes are inherited from the person's parents, they contain recombined segments of DNA from all a person's ancestors. Admixture testing compares an individual's DNA with specific sequences of DNA that are more common in people from one area of the world than another. Therefore, admixture testing allows us to determine which of the major biogeographical groups a person fits into—sub-Saharan African, European, East Asian, or Native American. The test results provide a percentage breakdown and can tell us something about many people, but they are not complete. Some major groups, such as African and South and Central Asian, are less represented in existing databases, but this is being improved on all the time. In addition, European groups are being further refined. The world has a lot of people on it, and when you split them into four populations, we discover that there are a lot we know little about. This has been made more complicated over the millennia due to the rise and fall of large empires and the spread of large global religions, which have played important roles in mixing tribes and ethnic groups.

(Answers to questions at end of the book—check each response before proceeding.)

1. Normally, a *sperm* will receive which of the following combinations of a particular chromosome pair?
   A. Both maternal and paternal
   B. Paternal only
   C. Maternal only
   D. Either maternal or paternal
2. If a *gamete* receives a paternal chromosome of chromosome pair A, which chromosome of pair D will it receive?
   A. Maternal
   B. Paternal
   C. Either maternal or paternal
   D. Neither maternal nor paternal
3. The principle illustrated in question 1 is an example of?
   A. Segregation
   B. Independent assortment
   C. Both
   D. Neither
4. The principle illustrated in question 2 is an example of?
   A. Segregation
   B. Independent assortment
   C. Both
   D. Neither

5. Is segregation mandatory or optional?
6. Is independent assortment a random or fixed procedure?
7. Preservation of the specific characteristics in humans is mainly a product of (segregation and independent assortment)?
8. Individual variation is mainly a product of (segregation and independent assortment)?

2.2 Self-Learning Cycle #2: Gametogenesis

Now, let's have a look at **Figure 2.4*** to see just how these principles work in humans. Actually, **Figure 2.10** represents a hypothetical animal with six chromosomes (three pairs), which makes it a little easier to understand, but the principles are the same and can be easily extended to any animal, including humans. The left side represents the formation of *sperm*, and the right side the formation of *eggs*. The genetic principles are identical, the main difference being that from each *male* cell that enters meiosis, four sperms are formed, while from each *female* cell entering meiosis, only one mature egg is produced. In the female, the other three (polar bodies) get practically no cytoplasm and eventually degenerate. Nevertheless, the genetic chance for a particular combination of maternal and paternal chromosomes in a particular egg is *the same* as it is for a particular sperm.

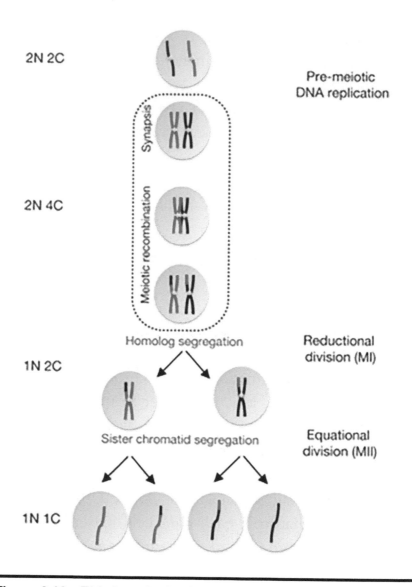

Figure 2.10 This is a single chromosome as it appears at metaphase. It consists of two daughter chromatids.

In the male, the entire process of germ cell formation occurs in the testis and is called *spermatogenesis*.

Spermatogenesis is divided into three phases, only *one* of which is termed *meiosis*. First, there is *mitotic* division of primordial germ cells, called *spermatogonia*, which leads to an increase in the number of spermatogonia, each with the typical *diploid* chromosome number (46 in man). This is represented in the top third of the figure and in the male is called *spermatocytogenesis*. When a particular spermatogonium stops multiplying by mitosis (after an undetermined number of divisions), it enlarges and matures into a primary *spermatocyte*. This represents the beginning of *meiosis* in the male. *Two* meiotic divisions ensue, during which segregation and independent assortment occur, resulting in the formation of four immature germ cells (spermatids) from each primary spermatocyte. Each spermatocyte has a haploid number of chromosomes, but with only two different combinations of maternal and paternal chromosomes out of the possible eight (independent assortment).

All eight possibilities will be realized in the meiosis of *all* the primary spermatocytes, but only two of these are possible in the meiosis of a single spermatocyte. (If crossing over, which we have not yet discussed, occurs, there will be *four* genetic variations.)

The prophase of the *first meiotic* division is relatively long, and several very important things happen, which are quite different from that which occurs in a typical mitotic division (**Figure 2.11**). First, each chromosome replicates during interphase (as in mitosis), but the *daughter chromatids* remain attached at the kinetochore.

Next, the maternal and paternal chromosomes of each pair of homologous chromosomes line up alongside one another, in a precise gene-to-gene pairing, called *synapsis*. This results in a four-member grouping of chromatids, known as a *bivalent*, or *tetrad*. There are 23 of these

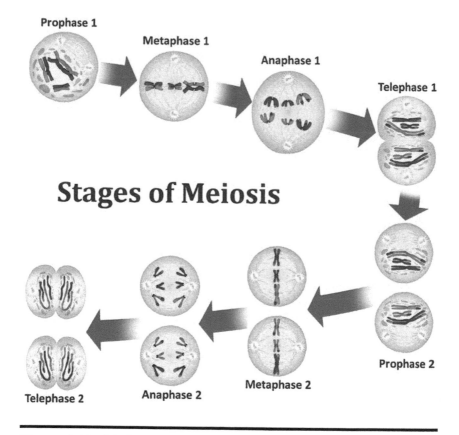

Figure 2.11   Various phases of meiosis.

bivalents in each primary spermatocyte, each consisting of a *pair of homologous chromosomes.* Each of the two chromosomes is replicated, but the two sister chromatids are still joined at the kinetochore. Thus, each bivalent has four chromatids or two pairs of sister chromatids (**Figure 2.10**).

The first meiotic division is unique and is typified by the kinetochores remaining functionally intact. (In mitosis, the kinetochore region splits so that the daughter chromatids can move to opposite poles, and thus to different daughter cells.) The failure of the kinetochore regions to divide in meiosis I means instead that *homologous chromosomes* separate and move to opposite poles and into different cells, achieving *segregation.* Sister chromatids of the same chromosome remain together. This first meiotic division, because of the segregation of the two chromosomes of a homologous pair, is often referred to as a *reduction division.* The two resultant cells of this division are called *secondary spermatocytes,* and as you can see from the figure, each contains different combinations of maternal and paternal chromosomes (independent assortment).

The *second* meiotic division can be considered a typical *mitotic* division (except that only half the chromosomes are present, to begin with), in which the kinetochores become functionally separate and the sister *chromatids* move apart and into different cells. Meiosis II is commonly thought of as an *equational* division. Thus, from the division of a pair of *secondary* spermatocytes, four spermatids are formed. Note from the diagram, however, that there are only two genetic *types* of spermatids. Each spermatid contains the haploid number (n) of chromosomes (one representative of each homologous pair), but different spermatids contain different combinations of maternal and paternal chromosomes.

The *third* stage of spermatogenesis is the *maturation* or *differentiation* (no division involved) of a spermatid into a sperm (spermatozoan) and is called *spermiogenesis*. The critical *genetic* events occur during *meiosis*, however, and we'll concentrate our present efforts there.

As we discussed earlier, the meiotic process is *genetically* the same in the female as in the male, although there are certain basic differences in location, number of cells involved, etc. These differences are not important for this discussion, however, and will be deferred until later. It is only by chance that the ootid ended up with the combination of chromosomes that it did. Any of the genetic combinations in the polar bodies might have ended up in the egg.

Meiosis

Homologous Chromosomes   Synapsis   Crossing Over

Figure 2.12   Gene crossover and synopsis.

Please test your knowledge before going further.

9. In the division of which of the following do sister chromatids separate and move to opposite poles? *(Multiple responses possible.)*
   A. Spermatogonium
   B. Primary spermatocyte
   C. Secondary spermatocyte
   D. Spermatid
10. What is the total number of *chromatids* found in a primary spermatocyte of a man?
11. How many *chromosomes* are represented in a primary spermatocyte of a man?
12. How many kinetochore regions are represented in a primary spermatocyte of a man? *(Multiple responses possible.)*
    A. Spermatogonium
    B. Primary spermatocyte
    C. Secondary spermatocyte
    D. Spermatid
    E. Spermatozoan
13. It is common to think of this cell as the *result* of a reduction division.
14. Independent assortment occurs during the division of this cell.
15. If you consider every chromatid a potential chromosome, which cell contains the haploid number chromosomes?

16. The division of which cell is commonly thought of as equational? Why?
17. How many sperm are produced from each primary spermatocyte?
18. How many genetic *types* of sperm are produced from each primary spermatocyte?
19. How many divisions are involved in spermiogenesis?
20. How many divisions are involved in spermatogenesis?
21. How many mature eggs are derived from a single primary oocyte?
22. Is the number of possible genetic combinations of eggs *less than, equal to,* or *more than* the number for sperm?

2.3 Self-Learning Cycle #3: A Simple Problem in Genetic Counseling

Although there is considerably more to meiosis than what we have discussed (we have thus far completely ignored *crossing over,* as some of you may have noticed), you may be surprised to learn that you now have enough information to counsel someone with a simple genetic problem, involving a single pair of genes on a pair of homologous chromosomes. All that we really must add are a few basic statements about how certain combinations of genes produce certain characteristics. At a particular locus (place) on a pair of homologous chromosomes, the two genes represented may be alike, or different. *Alleles are variations of genes occupying the same locus on homologous chromosomes and which affect the same function differently.* A person possessing two identical genes at a particular locus is said to be *homozygous*, while one possessing two different genes (alleles) at that locus is *heterozygous*. Consider a normal gene (A) and its abnormal allele (a), which can produce a condition in humans known as *albinism*: a lack of pigmentation. Albinism is produced only when a person inherits two (a) genes, that is, when he is homozygous for the gene (a). This is an example of *recessive* inheritance (two genes required for the expression of a trait).

Now, let's try to put it all together—segregation, independent assortment, recessive inheritance of a disease or condition—by following the results of a mating between two parents *heterozygous* for albinism. Neither has the condition because it takes two (a) genes to produce it. But is it possible for them to have children with this inherited disorder? Remembering that the genes in question are carried on a pair of chromosomes which must undergo meiosis, and thus follow the rules of segregation and independent assortment, let's diagram the possibilities for their offspring (**Figure 2.13a, 13b** and **13c**).

THE CHANCE OF FERTILIZATION OF A PARTICULAR EGG BY A PARTICULAR SPERM CAN BE REPRESENTED IN A BLOCK DIAGRAM AS FOLLOWS:

Thus, offspring can be expected in the following proportions:

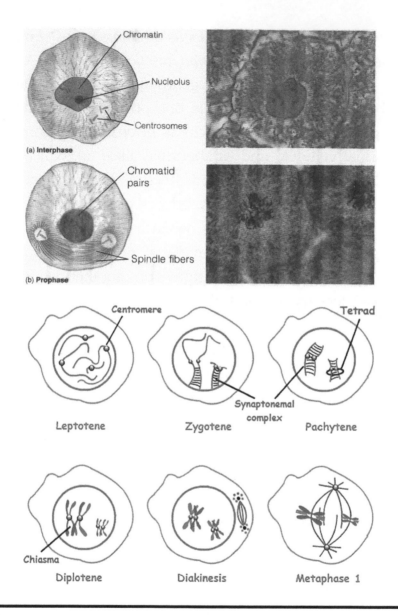

Figure 2.13a, 13b, and 13c   Process and stages of gene exchange. Top: electron micrograph of spindle formation (top) and illustration of various stages of prophase in meiosis. Middle: illustration of various stages of the gene exchange process. Bottom: the concept of homozygous and heterozygous alleles is illustrated.

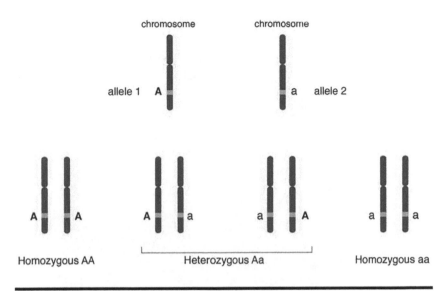

Figure 2.13a, 13b, and 13c   (Continued)

1/4 AA, completely normal (unaffected)
1/2 Aa, without albinism (unaffected), but carriers of
the defective gene (like the parents)
1/2 aa, albinism

Actually, the family would have to be very large for
these ratios to be achieved. A more realistic way of looking at it is that the fractions represent the *odds* for the
birth of *each* child; thus, there is a one in four chance
that each child will be an albino, and there are three
chances out of four that each child will be free of the
condition. Of those free of albinism, however, there is a
2/3 chance that they are carriers (Aa) (**Figure 2.14a** and
2.14b).

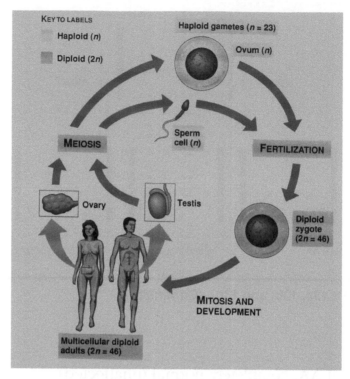

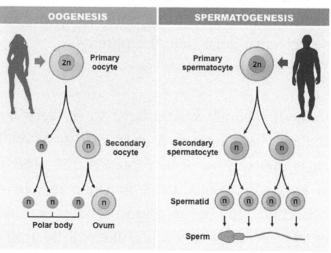

Figure 2.14a and 14b **The process of haploid DNA formation in female ovaries and male testicle.**

2.4 Self-Learning Cycle #4: Crossing Over

Not too long ago we discussed, in some detail, the major events that occur during the two meiotic divisions of male and female germ cells: *segregation* and *independent assortment*, or combining the two, *independent segregation*.

Another event of significance that occurs during the *prophase* of the *first* meiotic division (that is, during the prophase in *primary* spermatocytes and oocytes) is genetic *crossing over*. Basically, this involves a physical exchange of genetic material between homologous chromosomes and has the consequence of effectively adding an unknown, but quantitatively important, degree of genetic variation to that already supplied by independent assortment. It lessens even further the chance that any two sperm or eggs from the same individual will have the same group of genes (genome), and of course, that principle extends to any two fertilized eggs produced by the same couple.

24. Name two meiotic phenomena that provide for genetic diversity.
25. What is involved in crossover?

Figures 2.4 and **2.6** show in more detail the events that occur during meiosis. We shall concern ourselves with the events that comprise the *prophase* of meiosis I, those indicated by the asterisks in **Figure 2.14**: synapsis, crossing over, and the synaptonemal complex.

Figure 2.5 begins just prior to that stage of **Figure 2.4**, which marks the beginning of meiosis. The primary spermatocytes and oocytes of **Figure 2.14b** have already formed bivalents (tetrads), and the first transitional steps between gonium and primary "cyte" have been skipped.

The *interphase* preceding prophase of meiosis I is characterized by little or no chromosomal movement and, as in mitosis, replication of the cell's DNA. The *prophase* of meiosis I is commonly divided into five stages: leptotene, zygotene, pachytene, diplotene, and diakinesis.

As the cell enters prophase, the following significant events are observed:

Leptotene ("thin ribbon"). Chromosomes long, unwound, unpaired; each chromosome appears to be a single thread, even though its DNA has been replicated; individual chromosomes are not yet identifiable; nucleolus present.

Zygotene ("joined ribbon"). Chromosomes becoming condensed; synapsis (pairing) begins either at the chromosome ends (which may be attached to the nuclear envelope), at the kinetochore, or even elsewhere and proceeds in zipper-like fashion, toward both ends; in this manner, bivalents are formed; this unique, highly specific behavior is not displayed elsewhere; it is of the greatest importance to the success of sexual reproduction and, thus, the perpetuation of the species.

Pachytene ("thick ribbon"). Begins when pairing is complete; chromosomes short and thick; each bivalent now seen to consist of four parts (tetrad); eu- and heterochromatic regions of chromosomes visible; *chiasmata*, points of physical exchange between homologous chromosomes, are first seen.

Diplotene ("double ribbon"). Centromeric regions of homologous chromosomes begin to repel each other (*repulsion*), making the chiasmata easily visible; *terminalization* begins (in this process, the chiasmata seem to slide toward the chromosome ends as the homologues continue to separate).

Diakinesis ("movement apart"). The bivalents move to the periphery of the nucleus; nucleolus disappears; nuclear envelope disappears; terminalization is completed; the cell is ready for metaphase I.

At birth in the human, all oocytes are found in a sort of modified diplotene, the "dictyotene" or "dictyate" stage, and remain arrested there, some 12 to 50 years, until they either degenerate or proceed to ovulation! No new eggs are formed in postnatal life.

Now check your knowlege before going further.

26. In meiosis, when is DNA replicated?
27. Synapsis begins at?
28. Each chromosome appears double when DNA synthesis is complete? (True or false.)

29. Define chiasmata.
30. Chiasmata are first seen at?
31. Chiasmata are first seen clearly during _____ due to?
32. The events leptotene through diakinesis occur during?
33. Arrange the stages between leptotene and diakinesis in their proper order.
34. In what stage are human oocytes at birth? Where does this fit into the meiotic scheme?

The crossing-over phenomenon occurs during the prophase we have just described. The chiasmata are visible manifestations of this genetic exchange and mark the approximate point at which crossing over has occurred (before criminalization, of course). Now, let's examine the details of crossing over and its genetic consequences.

Figure 2.13 begins with a single pair of chromosomes as they might appear at diplotene. In the top part of the diagram (A), the bivalent can be followed through meiosis I and II, in the absence of a crossover. At meiosis I, segregation has occurred for all parts of the chromosomes, because maternal and paternal components have moved as units into different secondary "cytes." For reasons that will become apparent soon, this is the only condition (no crossing over) under which a complete reduction division can occur during meiosis I. To repeat, for this is an important point that is often misunderstood, *reduction* means the separation of maternal and paternal genes into different germ cells, this has occurred completely at meiosis I.

Now, turn your attention to the lower part of **Figure 2.12** and **Figure 2.13b** and note that a single crossover event, indicated by a single chiasma, has occurred and is observed at diplotene. By studying the diagram carefully, the following points should become clear:

■ Each crossover event occurs between only two of the four chromatids of a tetrad, specifically between individual chromatids of maternal and paternal chromosomes of the homologous pair.

■ The exchange is reciprocal.

■ For the segments of the chromatids that have been exchanged, reduction division has not occurred at meiosis I: maternal and paternal genes have not separated into different germ cells.

■ For these exchanged segments, reduction division occurs at meiosis II; for those segments not exchanged, at meiosis I.

■ Among the four products of meiosis II, the spermatids and ootids, there are two "parental" type chromosomes and two "recombinants"; important use is made of this fact in linkage studies in animals, which are used to determine how far apart genes on the same chromosome are and to construct chromosome maps.

■ If the genes at a particular locus involved in a cross-over are identical (homozygous condition), no effect will be observed; if they are different (heterozygous), the genome will have been shuffled, adding to the genetic diversity of the gametes.

35. What part of a chromosome will always undergo reduction division at meiosis I?
36. How many tetrads of a single bivalent are involved in a crossover event? Which one(s)?
37. Meiosis I can be uniformly characterized as a reduction division? (True or false.)
38. Exchanged segments of homologous chromosomes segregate at meiosis II? (True or false.)

It is important to note that the frequency that any two genes on the same chromosome are separated by a cross-over is proportional to the distance they are apart on the chromosome; this will show up in the frequencies with which parental and recombinant gametes are produced when two pairs of genes are tested against one another. Thus, if 80% of the gametes turn out to be parental types and only 20% recombinant, the two genes in question are assumed to be relatively that much closer than when one of the genes is tested against a third and the number of recombinants turns out to be, say, 30%. In this way, relative distances of different genes on a chromosome can be plotted. Since the necessary test crosses cannot be made in humans, other, more sophisticated techniques, based on cell hybridization, are being employed, but this exciting phase of genetics will have to be deferred until later.

One interesting consequence of the crossover phenomenon is that, as genes move farther apart on the chromosomes, they become separated from one another regularly. Thus, crossover becomes the rule rather than the exception, and when the number of recombinant gametes equals the number of parental gametes, that is, when there are equal proportions of all four gametes resulting from the meiosis of a primary "cyte," it cannot be determined whether the two genes are far apart on the same chromosome or whether they are on different chromosomes. In both cases, the two sets of genes show independent assortment!

This may become clearer by demonstrating just how the test backcross works in lab studies. Suppose two genes, in a doubly heterozygous state, are to be tested against one another, say, genes B and C of **Figure 2.6b**. If they are on different chromosomes, they will, of course, assort independently. The gametes formed, because of independent segregation, will be BC, BC', B'C, and B'C'. If this individual is mated to a doubly homozygous individual, say, B'B'C'C', only B'C' gametes will be formed by the latter, and the offspring will be of four different types, in equal proportions:

BB'CC', BB'C'C', B'B'CC', and B'B'C'C'. By the same token, if the genes are linked, that is, on the same chromosome, but so far apart that they are regularly separated by a crossover event, the gametes formed will be those shown in **Figures 2.11, 2.13b,** and they will be formed in equal proportions. It should be obvious that the result of a test backcross as stated previously will yield the same offspring as if the genes were on different chromosomes! Again, this illustrates the regularity of crossing over.

The in-between case, in which linked genes are close enough not to be separated regularly by crossover, is the one used to give linkage data, and in this case, the gametes will be the same. However, there will be fewer recombinant gametes than parental types, and these will show up in a test backcross as fewer recombinant offspring. If the two linked genes are lightly linked, that is, close enough that they are rarely, if ever, separated by crossover events, then only the parental types will emerge.

Such is the nature of what occurs during the prophase of meiosis I; its consequences, you should be convinced by now, are to further reduce the possibility that offspring of the same parents, indeed of the whole population, will be alike.

39. How would you describe a test backcross?
40. What kind of information is derived from a test backcross?

41. If two pairs of genes assort independently, what conclusion may be drawn?
42. What conclusion can be drawn from the fact that genes far apart on the same chromosome assort independently?

2.5 Self-Learning Cycle #5: Synaptonemal Complex

During *zygotene*, when the homologous chromosomes are close together (within 100–300 nm), a structure designated the *synaptonemal complex* develops between them. In most organisms, genetic recombination (crossing over) takes place while the complex is present and seems to be necessary for meiotic recombination. Some organisms in which chiasmata are absent (no crossing over) lack synaptonemal complexes.

The synaptonemal complex is seen at the ultrastructural level as a pair of dark parallel ribbons separated by a less-dense space. It has several components which may have somewhat-different substructures in different organisms, but the basic components seem to be homologous.

The complex itself, which consists of two *lateral elements*, *transverse filaments*, and a *central element*, is located axially to a pair of homologous chromosomes.

The *lateral element* develops from an *axil core*, or *axial element*, which is a densely staining ribbon seen associated with the *unpaired* chromosomes at leptotene or later stages. The origin of this ribbon is uncertain, but it is closely associated with the chromatin of the chromosome, and as the homologues approach one another closely, it becomes the *lateral element* of the synaptonemal complex.

The *central element* is a ladder-like configuration in the center of the complex. There is some evidence that it arises in the nucleolus.

The *transverse filaments* interconnect the central element with the lateral elements. They are arranged in sheets, and it has been suggested, on the basis of their varying lengths at the bifurcations that exist while synapsis is in progress, that they may be contractile, aiding in the final precise approximation of the homologues (gene-to-gene pairing).

The exact chemical nature of the components of the synaptonemal complex is uncertain, but it now seems that they are proteins. They seem to be assembled for their specific purpose, after which they are discarded. From the previous text, it appears that the separate elements of the complex may even be derived independently and assembled later.

43. What is the apparent basic function of the synaptonemal complex?

44. Which component of the synaptonemal complex develops in close approximation to the chromosome, and what is its fate?
45. What is the apparent role of the nucleolus in the formation of the synaptonemal complex?
46. For the sake of review, what other functions have been assigned to the nucleolus?
47. What relationship do the chromosomes bear to the nuclear envelope?
48. What role is proposed for the transverse filaments?

Exam 1: *Genetics*

A. *Mitosis*   **B.** *Meiosis*   **C.** *Both*   **D.** *Neither*

1. Process found in germ cells of the testis.
2. Process limited to germ cells.
3. Pairing of homologous chromosomes is seen.
4. Reduction division occurs.
5. Occurs in the male testis throughout his life.
6. Genetic variation is a consequence of _____?
7. The haploid number is produced during this process.

A. *Segregation*   **B.** *Independent assortment*
C. *Both*   **D.** *Neither*

8. Refers to the exchange of segments of homologous chromosomes.
9. Occurs during meiosis.

10. Refers to the separation of homologous chromosomes into different cells.

11. Refers to the way different pairs of chromosomes are distributed (relative to one another) to germ cells.

12. The process commonly considered to result in a "reduction" division.

13. How many viable gametes are eventually produced by complete meiosis of one primary spermatocyte?

    A. 1   B. 2   C. 3   D. 4

14. Applying the principles of meiosis, how would you counsel parents of one child with sickle cell anemia, knowing that it is a recessive disease, as to the chance of their next child having the disease? Both parents are normal, that is, free of that disease. The next child and each subsequent child would have:

    A. 25% chance   B. 50% chance   C. 0% chance

15. How many new eggs are produced during the post-natal life of a female?

    A. 0   B. Millions   C. Thousands   D. Few hundred

16. Prior to the time an oocyte begins its preparation for ovulation, which of the following best describes its condition?

    A. Has completed meiosis
    B. May be in any stage of meiosis
    C. Is in an arrested prophase
    D. Is in a premeiotic stage

A. *Independent assortment*   **B.** *Crossing over*
C. *Both*   **D.** *Neither*

17. Refers to the separation of maternal genes.
18. May occur at meiosis II.
19. Contributes to genetic variability.
20. Occurs when the synaptonemal complex is present.
21. Occurs during prophase of meiosis I.

A. *Zygotene*   **B.** *Diplotene*
C. *Both*   **D.** *Neither*

22. Repulsion occurs.
23. Chiasmata are seen.
24. Synapsis begins.
25. Tetrads are observed.
26. Marks the end of prophase.

A. **DNA**   B. **RNA**
C. **Protein**   D. **B and C**   E. **All**

27. Gene expression consists of?
28. Gene storage in mammals.
29. Gene storage in retroviruses and flu virus.
30. Found in the nucleus.
31. Found in the cytoplasm of mammalian cells.

A. **Probe**   B. **Primer**
C. **Both**   D. **Neither**

32. DNA.

33. RNA.
33. Is used to detect a gene of interest.
34. Is used to initiate a gene replication.
35. Restriction endonucleases are especially useful if they generate sticky ends (or hanging ends).
    What makes an end sticky?
    A. Single-stranded complementary tails
    B. Blunt ends
    C. Poly A sequences
    D. 5' cap

    A. Ligase   B. Plasmid
    C. Antibiotic resistance   D. cloning sites
    E. YAC
36. Is utilized to fill in the gap between the two DNA strands.
37. Is utilized to grow the recombinant gene plasmids.
38. Are utilized to insert the desired DNA into the plasmids.
39. Basic tool of cloning a small-size gene.
40. A way to clone a large-size gene.
    A. Okazaki fragment   B. Ligase
    C. Leading strand   D. Lagging strand
    E. A and B
41. Fast synthesis of DNA.
42. Slower DNA replication.
43. Found in which strand.
    A. Template strand   B. Sense strand
    C. Ligase enzyme   D. non-template strand

44. RNA transcription.
45. Same sequence as RNA.
46. Template strand.
47. Initiation codon ATG is located.
48. TATA box located.
49. Promotor is located in . . . ?
50. Seals DNA fragments.
51. Alternate forms of a gene that influence the same trait and are found at the same locus in homologous chromosome are called:
    A. Alleles
    B. Phenotype
    C. Genotype
    D. Heterozygous
    E. Homozygous
52. Which of the following is represented by word description, such as black and tall?
    A. Phenotype
    B. Genotype
    C. Both
    D. Neither
53. In humans, brown eyes are dominant over blue eyes, a brown-eyed (B) woman who has a blue-eyed (b) child has the genotype?
    A. bb
    B. Bb
    C. BB

D. All of the above

E. None of the above

Answers: Practice Cycles

1. D
2. C
3. A
4. B
5. Mandatory, in a normal meiotic division
6. Random
7. Segregation
8. Independent assortment
9. A, C
10. 92
11. 46 (each one, however, is replicated; there are 92 *potential* chromosomes)
12. 46
13. C
14. B
15. D, E
16. A, C, because kinetochores divide and sister chromatids move to opposite poles; the second meiotic division is like an ordinary mitotic division (e.g., spermatogonium) in this respect.
17. 4
18. 2 (If you already know about crossing over, which we have not yet discussed, you may have answered 4.

If that's what you had in mind, your answer is OK.)

19. 0
20. Undetermined because number of spermatogonial divisions unknown
21. 1
22. Equal to
23. a.
    b. Refer to diagram on page x—the diagramming is identical.
    c. 1/4
    d. Segregation. When only one pair of genes is involved, segregation is the primary concern. Independent assortment comes into play in considering the inheritance of two or more pairs of genes, especially when they are on different chromosomal pairs.
24. Independent assortment and crossing over.
25. Physical exchange between homologous chromosomes.
26. During interphase, as in mitosis.
27. Zygotene.
28. F, during pachytene.
29. Chiasmata (plural)—points at which homologous chromosomes (chromatids) cross over one another and which mark sites of physical exchange of genetic material.
30. Pachytene.

31. Diplotene, repulsion of centromeres.
32. Prophase of meiosis I.
33. Zygotene, pachytene, diplotene.
34. Dictyate, or dictyotene; this is a modified diplotene.
35. Centromere (kinetochore).
36. Two; two nonsister chromatids.
37. F.
38. T.
39. Doubly heterozygous vs. doubly homozygous individuals.
40. Relative distance genes are apart on chromosome from % of recombinant offspring.
41. They are either far apart on same chromosome or on different chromosomes.
42. Crossover is the rule rather than the exception.
43. The central element seems to develop within it.
44. Synthesis of rRNA, assembly of ribosomal subunits, necessary for movement of RNA to cytoplasm.
45. Both ends are attached, at least during prophase of meiosis I, to the nuclear envelope, apparently by the axial cores (lateral elements).
46. Contractile.

Chapter 3

Structure of DNA and Why We Use DNA in Forensic Science

What Are Small Tandem Repeats (STRs), and Why Do We Use Them?

This chapter will focus on the structure of DNA and the reason that the analysis of DNA is one of the most important ways to identify a person. We will focus on the molecular biology of DNA fingerprinting. What is DNA fingerprinting? How do we analyze the DNA found at a crime scene? How is it used to find your ancestors, and how is it employed for paternity testing and forensics?

DNA can be used to tell people apart since humans differ from each other based on their DNA sequence or the lengths of repeated regions of DNA. Length differences in the repeated regions of DNA are generally used in forensics, in criminal investigations, and in paternity testing.

DOI: 10.1201/9781003182498-4

3.1 What Is DNA?

DNA, or deoxyribonucleic acid, is the hereditary material in humans and almost all other organisms. Nearly every cell in a person's body has the same DNA. Most DNA is found in the cell nucleus. In humans, there are 23 pairs of chromosomes, in which our nuclear DNA is located. We call this **nuclear DNA** (nDNA), and it is the most common DNA used in forensics and during paternity testing. But a small amount of DNA can also be found in the mitochondria, the so-called **mitochondrial DNA, or mtDNA** for short. We will learn a great deal about mitochondria later and how we can determine maternal ancestry using mtDNA. You already know from Chapter 1 that mitochondria are structures within cells that convert energy from food into energy (ATP), the primary source of cellular functions (**Figure 3.1**).

You already know that half of our DNA comes from our mother and the other half from our father. The information in DNA is stored as a genetic code made up of four chemical bases: adenine (A), guanine (G), cytosine (C), and thymine (T). Human DNA consists of about 6 billion bases, and more than 99% of these bases are the same in all people. The order, or sequence, of these bases determines the information available for building and maintaining an organism, like the way in which letters of the alphabet appear in a certain order to form words and sentences.

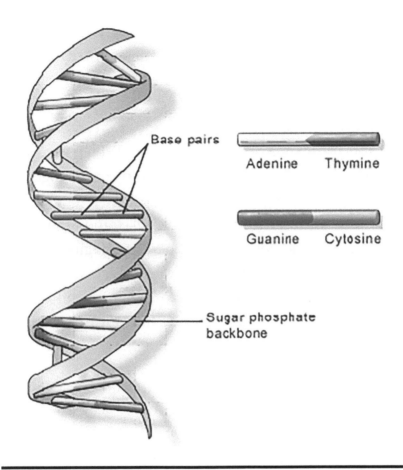

Base pairs

Adenine   Thymine

Guanine   Cytosine

Sugar phosphate
backbone

**Figure 3.1 DNA double helix. The blue outer structure is
made of a sugar and phosphate backbone. The inner base pair
sequence or order determines the uniqueness of each of us
as a human. The sequences of four base pairs—adenine, thymine, guanine, and cytosine—are four bases that are the most
important in DNA fingerprinting, for identification, paternity,
and ancestry determination.**

Source: US National Library of Medicine.

DNA bases pair up with each other, **A with T** and **C with G**, to form units called base pairs (bp). This **base pairing rule** is important since A-T and C-G are always together in the opposite strands of the double helix. Thus, if we know the sequence of one strand, we can easily figure out the sequence of the other strand. Each base is also attached to a sugar and phosphate molecule. Together, a base, sugar, and phosphate are called a nucleotide. Nucleotides are arranged in two long strands that form the famous double helix spiral. One can imagine that the structure of the double helix is like a ladder, with the bp forming the ladder's rungs and the sugar and phosphate molecules forming the ladder's side rail (**Figure 3.2**).

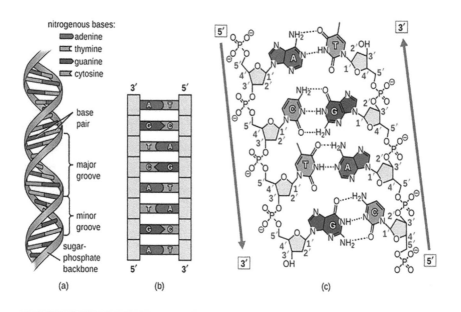

Figure 3.2 An overview of the DNA structure is depicted.

Source: DNA structure diagram, Google Search.

An important property of DNA is that it can make copies of itself: it can replicate. Each strand of DNA in the double helix can serve as a blueprint for duplicating the sequence of bases. This is critical when cells divide, because each new cell needs to have an exact copy of the DNA that was present in the old cell (**Figure 3.3**).

DNA's unique structure enables the molecule to copy itself during cell division. When a cell prepares to divide, the DNA helix splits down the middle and becomes two single strands. Each of the single strands serves as a template for building the two new opposite strands, creating double-stranded DNA molecules—each a replica of the original DNA molecule. Therefore, in this replication process, an **A** base is

DNA replication fork

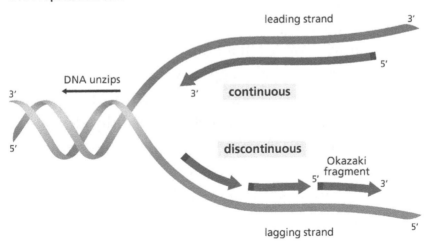

Figure 3.3 DNA replication.

Source: DNA Replication (Updated), YouTube. (DNA replication— 3D, YouTube.)

added wherever there is a **T**, a **C** where there is a **G**, until all the bases once again are paired. This pairing-based replication is called complementary strand synthesis. We can watch this whole replication process on YouTube.

The middle molecules: ATCG, and the order in which they appear, is the key to understanding how each of us is unique! It is also the key in understanding how and why we use DNA for forensic analyses. In **Figure 3.2**, the middle molecules are shown in four **colors**: thymine, adenine, guanine, and cytosine. These four different molecules are called nucleotides. The sequences of these four bases carry the instructions that are needed for any living being to survive and develop. In the previous section we discussed mitosis and meiosis. All the processes we discussed in the proceeding sections are governed by the sequences of these four nucleotides in a human (and all living organisms). A single fertilized egg in your mother's uterus is regulated by the sequences in the DNA of these four nucleotides: from a single cell into a fully developed newborn and then from a baby into a fully reproductive man or woman. Each DNA sequence that contains instructions to make proteins is known as a gene. The size of a gene may vary significantly. However, the sequences of DNA that make a gene represent only 1% of DNA; the other 99% are involved in regulating the genes, and some are of unknown function. Of note, all humans have similar DNA sequences, and 99% of them are similar! Then how do we differentiate one human from another by using DNA as a major tool for DNA-based analyses during crime scene investigations and paternity testing? Even though we do use the 1% DNA sequence difference, we also use an amazingly different method for crime scene investigation and for paternity testing. Let us first understand how we represent DNA sequences.

3.2 The Structure of DNA

5'! 3'

A	**! T!**
T	**! A!**
G	**! C!**
C	**! G!**
C	**! G!**
G	**! C**
A	**! T!**
T	**! A!**
C	**! G!**
G	**! C!**
T	**! A!**
T	**! A!**

3'! 5'!

Figure 3.4 shows DNA sequences in a vertical fashion. DNA sequences can also be shown in a linear fashion from the vertical format. We only show one side of the ladder of the double helix—meaning, in 5" to 3" order—since we know the other side will be the antiparallel-3'-to 5' and **T** will always pair with **A** and **C** will always pair with **G**.

From the previous image you may think that there are lots of letters, but remember, an individual human cell's

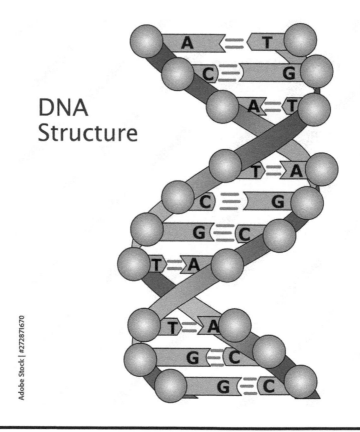

Figure 3.4   Representation of DNA sequence base pairs: from the double helix to vertical letters.

DNA contains **six billion bases**, three billion inherited from your mother, and three from your father!

Specific DNA sequences, on the same chromosome ONLY, differ at a few bases in everyone. However, all humans differ from each other in certain ways at the level of DNA, except for identical or maternal twins, whose DNA sequences are similar . . . or are they? (More in Appendix 1 on this subject.) Therefore, how do we identify one person from another (**Figure 3.5**)?

```
AAGTCAAGCTGCTCTGTGGGCTGTGATCTGCCTCAAACCCACAGCCTGGGTAGCAGG
AGGACCTTGATGCTCCTGGCACAGATGAGGAGAATCTCTCTTTTCTCCTGCTTGAAG
GACAGACATGACTTTGGATTTCCCCAGGAGGAGTTTGGCAACCAGTTCCAAAAGGCT
GAAACCATCCCTGTCCTCCATGAGATGATCCAGCAGATCTTCAATCTCTTCAGCACA
AAGGACTCATCTGCTGCTTGGGATGAGACCCTCCTAGACAAATTCTACACTGAACTC
TACCAGCAGCTGAATGACCTGGAAGCCTGTGTGATACAGGGGGTGGGGGTGACAGAG
ACTCCCCTGATGAAGGAGGACTCCATTCTGGCTGTGAGGAAATACTTCCAAAGAATC
ACTCTCTATCTGAAAGAGAAGAAATACAGCCCTTGTGCCTGGGAGGTTGTCAGAGCA
GAAATCATGAGATCTTTTTCTTTGTCAACAAACTTGCAAGAAAGTTTAAGAAGTAAG
GAATGA, TGTGATCTGCCTCAAACCCACAGCCTGGGTAGCAGGAGGACCTTGATGC
TCCTGGCACAGATGAGGAGAATCTCTCTTTTCTCCTGCTTGAAGGACAGACATGACT
TTGGATTTCCCCAGGAGGAGTTTGGCAACCAGTTCCAAAAGGCTGAAACCATCCCTG
TCCTCCATGAGATGATCCAGCAGATCTTCAATCTCTTCAGCACAAAGGACTCATCTG
CTGCTTGGGATGAGACCCTCCTAGACAAATTCTACACTGAACTCTACCAGCAGCTGA
ATGACCTGGAAGCCTGTGTGATACAGGGGGTGGGGGTGACAGAGACTCCCCTGATGA
AGGAGGACTCCATTCTGGCTGTGAGGAAATACTTCCAAAGAATCACTCTCTATCTGA
AAGAGAAGAAATACAGCCCTTGTGCCTGGGAGGTTGTCAGAGCAGAAATCATGAGAT
CTTTTTCTTTGTCAACAAACTTGCAAGAAAGTTTAAGAAGTAAGGAATGA and
```

Figure 3.5   Linear representation of a DNA sequence. It is a general practice to display the sequence of only one strand of the double helix, that is, 5' to 3'.

Source: https://en.wikipedia.org/wiki/Nucleic_acid_sequence.

3.3 DNA Fingerprinting or Genotyping

The method that we use to distinguish people is called DNA fingerprinting. It is an incorrect term, however, and we only use it as an analogy to the fingerprint technique employed by traditional forensics, that is, taking a print of the ridge patterns of fingers. The more precise term is *forensic genotyping*, and **Figure 3.6** will explain how this works. The illustration that follows shows a small DNA sequence of individual #1 and how it is written down. The top shows the top strain, representing the sequence in 5'-to-3' fashion, and the double helix of DNA is anti-parallel, meaning, it runs in the opposite direction, 3'-to-5', the bottom sequence.

The sequence for individual #2 may only differ to that of individual #1 by one nucleotide. So how do we differentiate one individual from another? Amazingly, there is a technique that can identify one individual with 100% precision! This forensic technique uses STRs, or **"short tandem repeats."**

Figure 3.6   An example of linear DNA sequence.

3.4 Short Tandem Repeats (STRs)

STRs, which are sometimes referred to as microsatellites or simple sequence repeats (SSRs), are accordion-like stretches of DNA containing core repeat units of between two and seven nucleotides in length that are tandemly repeated approximately a half dozen to several dozen times. Think of this as a colored rope where the length of the color band is repeated many times. This length can be different in different individuals. In the following example, the rope has two bands of red which repeat many times. Think of this as a repetitive DNA sequence with **ATATATAT** (four repeats) in one individual and the same repetitive sequence at the same location in individual 2 **ATATATATATAT** (six repeats). If we can separate DNA by size, it will be easy to tell the two individuals apart. This is carried out by a method called gel electrophoresis.

3.5 Homozygous and Heterozygous STR Sequences

Before we go to DNA strand separation by size, let us clarify an important fact so the STR technique is clearer.

In the proceeding section, we discussed homozygous and heterozygous genes. In the STR-based analysis of

DNA, we use the same idea. So a particular STR (i.e., **ATATATAT**) in an individual can be homozygous or heterozygous. For example, you might have inherited an **AT** STR from your mother as four repeats, while that from your father as eight repeats. In this case, you will be heterozygous 4/8 for STR **AT**. On the other hand, your best friend might have inherited the same STR (AT) from both of his parents as six-repeats: **ATATATATATAT.** He is then homozygous for AT STR. In another friend, the AT repeats, on the same chromosome, may be **4/4,** or homozygous, or it could be **8/16,** which is heterozygous (**Table 3.1**).

The human genome contains several thousand STR markers, but only a small core set of STRs loci has been selected for use in forensic DNA and paternity testing. However, to give the human identification database a universal commonality, all forensic laboratories use the same sets of STRs. There are slight differences in the numbers of STRs we use in the USA compared to what the European forensic laboratories use. We use a total of 13 STR, whereas the Europeans use a few more (14 or 16, depending on the country). The utilization of the same loci with the same STR repetitive sequences allows the whole globe to share the same core loci and permits genetic information to be shared and compared by utilizing highly confidential information. One of the key elements of utilizing the core STR loci is that it allows the use of commercial kits that are standardized and are now

Table 3.1 Characteristics of the 15 STR Loci Present in the Commercially Available Kit AmpFlSTR Identifiler

STR Loci	Chromosomal Location	Repeat Motif	Allele Range[a]	PCR Product Sizes in Identifiler Kit (dye label)
CSF1PO	5q33.1	TAGA	6–15	305–342 bp (6-FAM)
FGA	4q31.3	CTTT	17–51.2	215–355 bp (PET)
TH01	11p15.5	TCAT	4–13.3	163–202 bp (VIC)
TPOX	2p25.3	GAAT	6–13	222–250 bp (NED)
VWA	12p13.31	[TCTG] [TCTA]	11–24	155–207 bp (NED)
D3S1358	3p21.31	[TCTG] [TCTA]	12–19	112–140 bp (VIC)
D5S818	5q23.2	AGAT	7–16	134–172 bp (PET)
D7S820	7q21.11	GATA	6–15	255–291 bp (6-FAM)
D8S1179	8q24.13	[TCTA] [TCTG]	8–19	123–170 bp (6-FAM)
D13S317	13q31.1	TATC	8–15	217–245 bp (VIC)
D16S539	16q24.1	GATA	5–15	252–292 bp (VIC)
D18S51	18q21.33	AGAA	7–27	262–345 bp (NED)
D21S11	21q21.1	[TCTA] [TCTG]	24–38	185–239 bp (6-FAM)
D2S1338	2q35	[TGCC] [TTCC]	15–28	307–359 bp (VIC)
D19S433	19q12	AAGG	9–17.2	102–135 bp (NED)
Amelogenin (sex-typing)	Xp22.22 Yp11.2	Not applicable	Not applicable	X = 107 bp (PET) Y = 113 bp (PET)

Source: The 13 core STR loci used for the U.S. national DNA database are shown in bold font. See www.cstl.nist.gov/biotech/strbase/multiplx.htm for information on other commercially available STR kits.

Figure 3.7 STR illustrated. Repetitive color represents STRs.

available to generate DNA profiles containing core STR loci. Millions of STR profiles are generated worldwide each year by the government, by universities, and by private laboratories performing various forms of human identity testing, including DNA databasing, forensic casework, missing persons/mass disaster victim identification, and parentage testing (**Figure 3.7**).

3.6 What Is the Gold Standard for Carrying Out DNA Profiling?

Is there a way to determine what kind of DNA profile a person has and how this can be assessed?

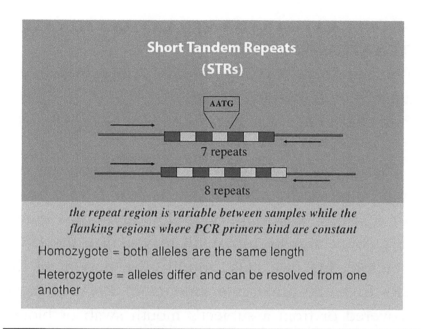

Figure 3.8 In this example, human #1 and human #2 have the same DNA sequence on a specific locus, but the lengths of their STR (i.e., GT) are different: #1 has 7 repeats, and #2 has 8 repeats.

Suppose two individuals have the same DNA sequence at a specific location on their chromosomes but their STR has different lengths. The STR **AATG** is the same between two individuals, but the DNA STR lengths are different (**Figure 3.8**).

Human #1

5'-ATGCC**AATGAATGAATGAATGAATGAATGAAT GA**GATCGTT-3'

Human #2

5'-ATGCC**AATGAATGAATGAATGAATGAATGA ATGAATGA**GATCGTT-3'!

In human #1 the AATG is repeated 7 times whereas in Human #2 it is repeated 8 times.

(short tandem repeat sequences profiler, Google Search)

That is, the length of their DNA is a little bit different from each other. If we isolate DNA from a hair or a tiny bloodstain on a cloth at a crime scene, the amount may be too small to visualize the DNA. Therefore, first, we must amplify the amount of the STRs from the minuscule amount of DNA we have recovered or from a subject's mouth swab or blood. Therefore, before we carry out gel electrophoresis, we must amplify the parts of the DNA that are in the STR regions of the DNA. Then, we can visualize the length difference by gel electrophoresis. To make hundreds of thousands or millions of copies of the specific STR, we use a technique called polymerase chain reaction, or simply **PCR**. A full picture of the process can be visualized by going to any of the sites on YouTube videos (i.e., https://online.universita.zanichelli.it/hillis-files/hillis_activity/act_1302_polymerase_chain_reaction/act_1302_polymerase_chain_reaction.html#). This site will allow you to visualize and learn the basics of PCR.

3.7 The Polymerase Chain Reaction (PCR)

The reaction is first heated to 95°C to melt (denature) the dsDNA into separate strands. The reaction is then cooled to ~50°C, at which temperature the primers will find base-complementary regions in the ssDNA, to which they will stick (anneal). The reaction is finally heated to 72°C, at which temperature the *Taq* enzyme (known as polymerase) replicates the primed ssDNA (extension). At the end of one cycle, the region between the two primers has been copied once, producing two copies of the original gene region. This is slightly oversimplified: for details, please watch the video on YouTube (www.youtube.com/watch?v=uKeMiAZ8Zu4).

Because a heat-resistant polymerase is used, the reaction can be repeated continuously without the addition of more enzymes. Each cycle *doubles* the copy number of the amplified gene: ten cycles ideally produce $2 \rightarrow 4 \rightarrow 8 \rightarrow 16 \rightarrow 32 \rightarrow 64 \rightarrow 128 \rightarrow 256 \rightarrow 512 \rightarrow 1{,}024$ (2^{10}) copies. Thus, 30 cycles yield a ($2^{10 \times 3} = 10^9$–fold amplification, or 10 billion copies of the original small STR sequence). PCR allows the production of a sufficient quantity of the STR of interest for direct analysis, for example, by gel electrophoresis (from PCR, mun.ca) (**Figure 3.9**).

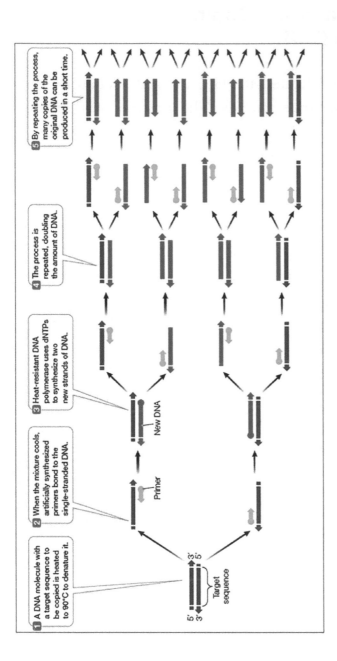

Figure 3.9 Briefly, PCR begins with a mixture containing a double-stranded (ds)DNA template either isolated at a crime scene or isolated from an individual, a pair of short single-stranded (ss)DNA oligonucleotide primers, a pool of the four deoxyribonucleotide triphosphates (dNTPs; ATCG), and a heat-resistant DNA polymerase, *Taq* enzyme. The reaction is carried out in a computer-regulated heating block—a thermal cycler, which permits rapid, controlled heating and cooling. The primers are chosen so that they are base-complementary to opposite ends of either strand of a short stretch of DNA containing the gene region of interest: PCR thus requires some prior knowledge of the gene sequence. In our case, the STR sequences.

When employing STR typing, PCR is used to recover information from tiny amounts of biological material (generally 1–50 nanogram). The relatively short PCR product sizes of approximately 100–500 base pairs (bp) generated with STR testing can be recovered even with degraded DNA that may be present due to environmental insults on the evidentiary biological material found at a crime scene. Because of the relatively short lengths of STR (generally in the range of four base pairs), PCR amplification of all STR (from 13 to 16) can simultaneously be amplified. This method, known as multiplexing, allows complete DNA fingerprinting in a single PCR run and, subsequently, the preparation of an individual's full DNA profile.

In summary, so far, we have covered what DNA STR-based forensic analysis is and how it can be used to identify individuals based on STR repeated lengths of DNA sequences. The STR length differences are generally used in forensics and paternity testing. The technique of gel electrophoresis separates DNA by size, and this allows the identification of people based on 13 STR lengths. If we use only one sequence of STR, **GATA,** for example, you may find one in ten individuals who may have the same GATA with the same numbers of repeats. And one in a ten would be exactly either homozygous or heterozygous for **GATA**. However, if we use 13 STRs, the chances that any two individuals will have the same DNA profile is more than one in ten billion.

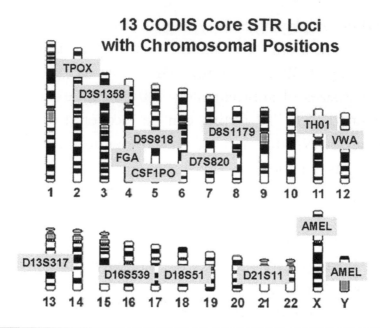

Figure 3.10 Positions of 13 STRs on human chromosomes.

Currently, 13 core STRs are approved by the FBI. Each of the STRs is mapped to a specific location on specific human chromosomes (**Figure 3.10**). Each of the STRs is designated by a name. For example, **TPOX** is located on chromosome #2 and is GAAT and generally repeats between 6 and 13 times. Mean GAAT can be between 24 and 52 bp in length.

3.8 Gel Electrophoresis

Now, let us explain the principle of the gel electrophoresis method. There are three simple facts we must understand first: (1) gel electrophoresis allows DNA fragments of different sizes to separate, (2) it uses a gel matrix that DNA molecules can move through, and (3) DNA is negatively charged.

The gel is generally made of agarose or polymers (**Figure 3.11**). One can imagine the gel matrix as a room full of springs going from one end of the wall to the other in 3D. If a bunch of people are asked to run from one end to the other end, it is obvious that the smallest among them will reach the other side first. Gel electrophoresis is similar, where we apply a mixture of DNA of various sizes at one end and then force them to migrate to the other end by using an electrical

Figure 3.11 Agarose gel polymer matrix.

current. Since DNA is negatively charged, we place the PCR-amplified DNA at the negative charge of the matrix, and the electric current flows through the matrix which is emersed in a buffer. The DNA rushes toward the positive charge. The smallest molecules reach the end first, while the larger molecules, of various sizes, migrate with different speeds. We use a stain that binds the DNA so that it can be visualized. If the DNA is bound to a fluorescent dye, we can visualize and photo-document the results. At the end of a run, the gel will look as shown in the following.

Let us summarize what we have learned so far and what is yet to come in the next chapter. For us to get the concepts solidified, let us use a crime scene to illustrate what is happening. Gel electrophoresis is a method that separates DNA of different sizes into precise bands. We use a standard DNA of known size and run an unknown sample by its side. After running the gel, unknown DNA sizes can then be easily calculated. On the left end of the gel, you can see we used a standard (markers) to visualize the known sizes of DNA. The three unknown samples of DNA display three different patterns. The three unknowns do not match each other. The graph on the right side shows the relation between DNA fragment size and distance traveled. The smaller they are, the farther they migrate (**Figure 3.12**).

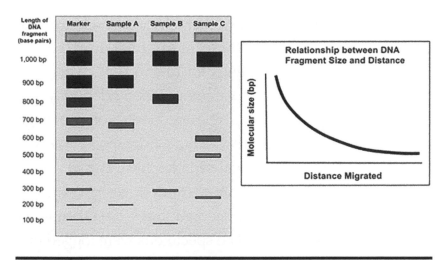

Figure 3.12 This figure illustrates how gel electrophoresis results are analyzed. The graph represents the relationship between DNA fragment size and distance traveled in the gel. The smaller the size of the fragment, the greater the distance traveled, as shown by the results on the left.

So can we use this technique to figure out the identification of an individual who might have committed a crime? The example that follows explains this. Let us go to the scene of a crime.

3.9 The Scene of the Crime

The body of a woman was found outside of an abandoned house and was accidentally discovered by mushroom hunters. After a 911 call, the police called in crime scene investigators (CSI). There were no bloody footprints or weapons found at the scene and no other obvious evidence that could lead to a killer. The CSI team carefully dusted down the body and its surroundings for fingerprints, hairs, and clothing fibers, possibly containing skin and bodily fluids. Importantly, they were able to collect some biological materials from under the victim's fingernails. The CSI team also collected biological samples from the dead victim's body to compare with the victim's fingernail samples. These materials were safely secured and were taken to an FBI-approved forensic lab. The specimens were logged into a secure computer and coded, so the specimens remained blind. Then the samples were thoroughly examined for biological evidence. Skin cells were found under the victim's fingernails probably belonging to the killer, indicating that a struggle took place. If the experts can isolate DNA from the skin cells, they can attempt to identify the killer using PCR and genetic fingerprinting using gel electrophoresis, thereby providing a lead toward solving the murder.

3.10 So Where Does PCR Come into All This?

In criminal investigations, the amount of DNA available for analysis is often limited to whatever can be isolated from a few skin cells, strands of hair, tiny amounts of blood, or other bodily fluids left behind at the scene. It can be very difficult to extract enough intact DNA from such material; thus, without PCR-based amplifications, where a minuscule amount of DNA can be used to make billions of copies of STRs, solving a crime scene murder, for example, would be difficult, indeed. Even with relatively degraded DNA, there is usually enough intact fragments that can be used to amplify sufficient STRs. PCR is a fast, cost-effective, and relatively easy solution to this problem, in that it can rapidly amplify specific STR sequences from the isolated DNA, increasing the amount of material and paving the way for further analysis.

3.11 How Is DNA Fingerprinting Actually Carried Out?

We use the DNA fingerprinting method by amplifying the known STR sequences shown in **Table 3.1**. This method takes advantage of highly polymorphic regions that have short repeated sequences, most often containing three to five repeated bases. Because unrelated people will almost definitely differ in their number of repeats, STRs can be used to confidently differentiate between unrelated persons.

3.12 Let Us See How Our Crime Scene Was Solved

DNA isolated from material collected at the crime scene was processed. The STR loci were amplified by PCR using sequence-specific primers. These primers are designed in such a way that we can amplify the STR loci from any individual. The resulting DNA fragments were then processed using gel electrophoresis. Separation and detection of these fragments result in a unique pattern of bands often referred to as a **DNA fingerprint**. It is this unique pattern that is ultimately used in criminal investigations to match suspect DNA with DNA found at the scene (**Figure 3.13**).

Summary of the Case: A genetic fingerprint can be directly used to match DNA found at a crime scene with suspect DNA to ultimately secure a criminal conviction. As with traditional fingerprinting, genetic fingerprinting

Figure 3.13 Results of the gel electrophoresis show the STR band patterns from three suspects. As noted previously, the gel pattern from the crime scene has a 100% match with suspect #2.

requires the presence of a corresponding fingerprint from any suspect under consideration to make an exact match. In other words, DNA collected at a crime scene must match actual suspect DNA that is already on file or developed during an investigation. From three potential suspects that emerged during the investigation, only one matched.

As well as using genetic fingerprinting to convict criminals, this technique also serves as a powerful tool to prove the innocence of suspects and previously wrongly convicted individuals. Alongside traditional fingerprint analysis, DNA fingerprinting is among the most unambiguous methods of identifying suspects available today. PCR has therefore revolutionized forensic science and criminal investigations, and in combination with traditional detective work, it will continue to be a powerful investigative tool in the future.

Chapter 4

Capillary Gel Electrophoresis to Analyze DNA in a Crime Lab

4.1 Automated DNA Sequencing by Capillary Electrophoresis

In Chapter 3 we learned how one can separate STR DNA fragments by gel electrophoresis. The major drawback of this technique is the need to prepare a fresh gel for each sequencing run. Each of the PCR products needs to be manually loaded into the wells. The manual loading of samples into the gel is cumbersome, and it takes a relatively long time for each run (i.e., 8 to 10 hours for resolution of a 1Kb DNA fragment) and has difficulties in the tracking of lanes. A minor mistake in loading the samples can result in an incorrect conviction and can spell disaster for some innocent person's life.

DOI: 10.1201/9781003182498-5

Instead of running gel in agar, an alternate and most used method for separating DNA fragments is capillary electrophoresis (CE) (**Figure 4.1**). During capillary electrophoresis, products of the sequencing reaction enter the capillary because of electrokinetic injection. A high-voltage charge applied to the buffered sequencing reaction forces the negatively charged DNA fragments into the capillaries. The DNA fragments are separated by size, with the larger fragments migrating more slowly through

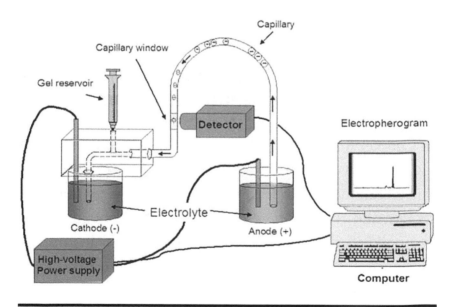

Figure 4.1 An overview of the capillary electrophoresis method. The amplified STR DNA fragments are loaded at the negative (cathode) end, and a high-voltage current separates DNA fragments according to their size. The information is then passed to a computer that creates an electropherogram.

the matrix. CE plays a central role in forensic labs across the globe in solving crimes and paternity testing.

Before reaching the positive electrode, the DNA fragments pass through a laser beam, which excites dye labels and cause them to fluoresce. The dye signals are separated by a diffraction system, and a charge-coupled device (CCD) camera detects the fluorescence. Since each dye emits light at a different wavelength when excited by the laser, all colors, and therefore loci, can be detected and distinguished in one capillary injection. The fluorescence signals are converted to digital data, in a file format compatible with an analysis software application. This instrument creates an electropherogram.

Let us summarize what we've learned so far about CE.

1) CE is an automated system—an instrument that is approved by the FBI. This highly computerized instrument is sold by Applied Biosystems (ABI).

2) In the CE instrument, STR DNA fragments are separated by size through a long, thin capillary filled with a polymer (usually polyacrylamide or acrylic fiber) instead of agarose gel that is used for gel electrophoresis.

3) A sample containing DNA fragments is injected into the capillary at the cathode end. A high-voltage electric current separates the DNA fragments according to their size. The very small diameter of the capillary,

and extremely high voltage, allows very rapid separation of DNA fragments.

4) During the PCR process, all DNA is labeled with four different colors, generating four different colors of STR DNA fragments. These four colors—green, blue, yellow, and red—are identified by a laser detector. We will learn how the colored DNA fragments are generated by PCR later in this book. This laser shoots through the capillary matrix, causing the colored tags on the DNA fragment to glow, each with a different wavelength.

5) The emitted colors of different wavelengths are detected by a camera and recorded by the machine.

6) The colors of the bases are then displayed on a computer as a graph of different peaks, as shown in the following (**Figure 4.2**).

In summary, most of the human genome is identical among diverse individuals. However, there are small

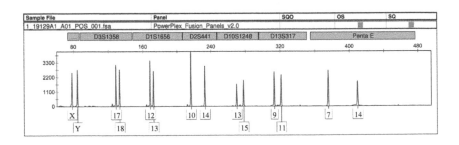

Figure 4.2 An example of the DNA profile of a male sample.

regions of variation, especially in areas with repetitive sequences. One example are short tandem repeats (STRs), which can vary in length among individuals, with each repeat consisting of three to five nucleotides of DNA. Analysis of these repeats reveals a characteristic pattern for everyone that is inherited from paternal and maternal DNA. Forensic DNA databases use these patterns of STRs for human identification, and each national database has requirements for specific STR loci that must be included.

To characterize these STR patterns or DNA profiles, human genomic DNA is amplified using the polymerase chain reaction (PCR) with sets of primers for each STR being tested. The primers are labeled with fluorescent dyes, and the amplified products are separated according to size by CE. Fluorescence imaging by the CE instrument results in a set of peaks for each dye channel, which together constitute an electropherogram that serves as a DNA "profile" for an individual. Statistical analysis of the DNA profile, and comparison to reference databases, can predict whether the individual matches a known DNA profile, with a high power of discrimination.

Similar in principle to gel electrophoresis, CE uses highly sophisticated equipment. These are computerized, automated, and complex machines that can do many analyses simultaneously and carry out the DNA profiling of dozens of DNA samples at a high speed and with precision.

CE uses different-colored fluorescent dyes for each of the 13 STRs and can separate as well as identify the sizes of each of the STRs, according to different-sized PCR products. The amplifications of multiple STR loci simultaneously in a simple reaction enable us to carry out a high power of discrimination in a single test without consuming much DNA (e.g., many times just 1 ng or less of starting material). We use this technology for crime investigations and paternity testing, and it is worth noting that these core STRs are in noncoding areas of DNA, meaning, they occur in between genes.

As we have discussed in Chapter 3, in the United States we use 13 markers or STR, and generally, it is necessary that all 13 STRs should exactly match the suscept. If one of the 13 STRs does not match or is in question, legally that individual cannot be indicted for the crime he/she is being accused of.

Chapter 5

Sex Determination by STR

Our first and most routine DNA biomarker is amelogenin. To be precise, we should state amelogenins, which are a group of protein isoforms coded by genes found in both male and female Y and X chromosomes and known as *AMELY* or *AMELX* genes (**Figure 5.1**). The *AMELY* gene is found in males, and two *AMELX* genes in females. Both genes code for proteins that have several isoforms in humans. Isoforms are any of two or more functionally similar proteins that have a comparable, but not identical, amino acid sequence. These isoforms are generally coded from the same gene, but during mRNA transcription, they change a bit due to posttranscriptional adjustments, where different exons have been removed by alternative splicing or proteolysis from the *AMELX* gene, on the X chromosome, and the *AMELY* gene in males, on the Y chromosome. They are involved in amelogenesis and the development of enamel. Enamel, you guessed it right, is the strongest biological material on the surface of our teeth.

DOI: 10.1201/9781003182498-6

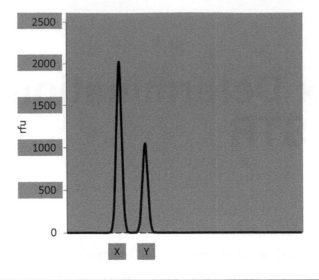

Figure 5.1 An electropherogram showing amelogenin locus for sex determination.

The amelogenin gene has been most widely studied in humans, where it is a single copy gene, located on the X and Y chromosomes at Xp22.1-Xp22.3 and Yp 11.2. It should be noted that both X and Y chromosomes have many genes besides amelogenin. X chromosomes have between 900 and 1,600 genes, whereas Y chromosomes have between 70 and 200 genes (**Figure 5.2**).

The differences between the X and Y chromosome versions of the amelogenin gene (*AMELX* and *AMELY*, respectively) enable it to be used in the sex determination of unknown human samples. *AMELX* is a six-base pair smaller than *AMELY*. To determine the sex from the DNA of an unknown, when we carry out PCR and use

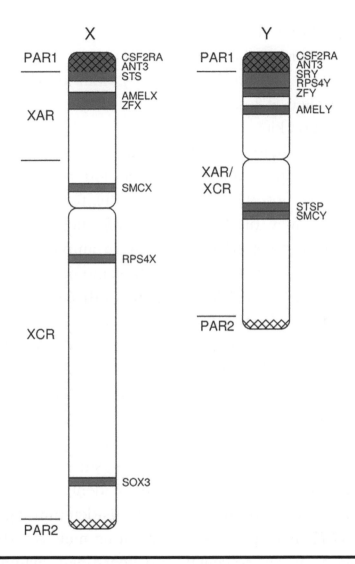

Figure 5.2 The illustration depicts the locations of various genes on X and Y chromosomes.

CE, we will recover "two bands" of DNA, at 106 bps for X and 112 bps for Y. If both the *AMELX* and *AMELY* versions of the gene are present, the sample is from a male, whereas if only the *AMELX* version is present, it is from a female. This can be accomplished using various amplification kits. The bp size may vary from kit to kit, but in the United States, we generally use the Profiler Plus Amplification kit.

Because there is *AMELY* variation among individuals and populations, this method of sex determination is not always accurate. Mutation in regions of *AMELY* may disable PCR amplification. Also, a 6 bp insertion to *AMELY* results in an amplicon identical in length to that of *AMELX*. In some males, *AMELY* may be completely deleted. In any of these cases, only one band is visualized during gel electrophoresis of PCR products, causing incorrect identification of the sample as female. The correct identification rate may vary among populations, but most of the time, the results are precise.

While in general, the amelogenin sex test is accurate, other Y chromosome markers, such as *SRY*, can be used for less-ambiguous gender identification. The *SKY* gene, also known as the "testis-determining factor" (*TDF*), is responsible for initiating male sex determination in all placental and marsupial mammals. Therefore, if not certain, we can use *SKY* as another marker to confirm the male origin of the DNA in question (**Figure 5.3**).

Sex Chromosomes

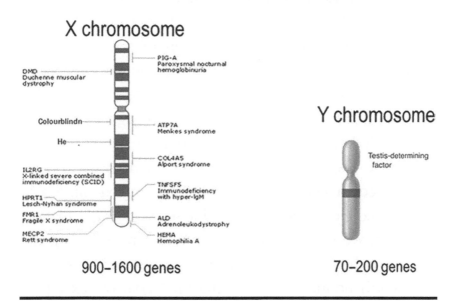

X chromosome

PIG-A
Paroxysmal nocturnal
hemoglobinuria

DMD
Duchenne muscular
dystrophy

Colourblindn

ATP7A
Menkes syndrome

He

COL4A5
Alport syndrome

IL2RG
X-linked severe combined
immunodeficiency (SCID)

TNFSF5
Immunodeficiency
with hyper-IgM

HPRT1
Lesch-Nyhan syndrome

FMR1
Fragile X syndrome

ALD
Adrenoleukodystrophy

MECP2
Rett syndrome

HEMA
Hemophilia A

900–1600 genes

Y chromosome

Testis-determining
factor

70–200 genes

Figure 5.3   The X and the Y chromosomes and their STRs.

5.1 Summary

In forensic casework, there is often a need to determine the sex of an individual based on DNA evidence. This is especially the case in instances such as victims of mass disasters, missing person investigations, and sexual assault cases. Currently, we use the amelogenin marker to identify sex from an unknown DNA or DNA from an individual suspected of committing a serious crime. This X-typing marker amelogenin is incorporated in most commercially available multiplex STR kits and is currently the most common sex typing marker used in forensic casework. For a DNA profile to be accepted into the US National DNA Index System (NDIS), which is the national level of the Combined DNA Index System (CODIS), the inclusion of the amelogenin marker is required for relatives of missing persons, along with the 13 CODIS core loci. The amelogenin test must be carried out for DNA profiles of missing persons and unidentified human remains. In addition to amelogenin, alternative Y-specific markers are also being used in place of or in conjunction with the standard amelogenin test to provide more consistent and reliable sex typing information. Currently, we use the *SKY* gene in addition to amelogenin in cases where sex determination results are ambiguous. It is worth noting that while the final goal of sex typing in forensic investigations is often to discover the legal or perceived gender of an individual, there will always exist cases in which an individual's chromosomal sex, as discovered by DNA sex typing, does not match legal gender or gender presentation. However, because legal gender and chromosomal sex do correspond in most cases, methods for accurate determination of chromosomal sex do and will continue to have widespread use in forensic investigations.

5.2 Learning Cycle 1: The X Chromosome Has More to It Than What Meets the Eye!

The genes in sex chromosomes X and Y have functions other than for reproduction, especially the X chromosome.

The X chromosome has hundreds of genes, and many of these regulate functions in organs other than those of the reproductive system.

- For example, a gene for color blindness present on X is responsible for the maintenance of photoreceptors in healthy cone cells of the retina.
- Then there are genes on the X chromosome which allow hepatic cells to produce clotting factors; hence, mutation in such X-linked genes causes hemophilia.
- The genes on the X chromosome are expressed not just in cells associated with the reproductive system; some of these genes are expressed in various somatic tissues of both sexes throughout life.
- On the other hand, the Y chromosome is very small and only present in males, that is, Y-linked genes are absent in all female cells.
- There is one *AMELY* gene on Y that codes for extra-cellular enamel protein amelogenin, which expresses during development of tooth enamel, a nonrepro-ductive tissue, but this is rather an exception.
- Most of the Y-linked genes are expressed in organs of the male reproductive system, such as testes, prostate gland, etc.
- Expression of the *SRY* gene of the Y chromosome initiates sex differentiation very early in embryonic life.

■ Therefore, expression of genes on the Y chromosome are restricted in male reproductive organs only, and some of these genes express for only a brief period of embryonic life.

Chapter 6

The Combined DNA Index System (CODIS)

DNA is present in nearly every cell of our bodies, and we shed our cells everywhere we go without even realizing it. Flakes of skin, drops of blood, hair, and saliva all contain DNA that can be used to identify us. In fact, the study of forensics, commonly used by law enforcement agencies around the globe, often relies upon these small bits of DNA to link criminals to the crimes they commit. This fascinating science is often portrayed on popular television shows as a simple, exact, and infallible method of finding a perpetrator and bringing him or her to justice (**Figure 6.1**). In truth, however, teasing out a DNA fingerprint and determining the likelihood of a match between a suspect and a crime scene is a complicated process that relies upon statistical probability to a greater extent than most people realize. In the

DOI: 10.1201/9781003182498-7

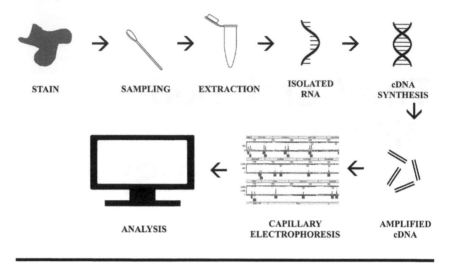

STAIN   SAMPLING   EXTRACTION   ISOLATED RNA   cDNA SYNTHESIS

ANALYSIS   CAPILLARY ELECTROPHORESIS   AMPLIFIED cDNA

Figure 6.1   Profiler Plus® kit results from a 1.2 mm FTA blood-stain disc (25-cycle amplification).

subsequent chapters, we will learn how to extract DNA from a tiny cell at the end of a hair or from under the bed, or from a small sliver of saliva left on a cigarette butt. Government-administered DNA databases, that is, CODIS, do help speed the process, but they also bring to light complex ethical issues involving the rights of victims and suspects alike. Thus, understanding the ways in which DNA evidence is obtained and analyzed, what this evidence can tell investigators, and how this evidence is used within the legal system, is critical to appreciating the truly ethical and legal impact of forensic genetics. We will discuss these issues in later chapters.

6.1 Use of STRs in Identification of Individuals

As we have discussed in Chapter 3, although most of the human genome is identical across all people, regardless of race or gender, there are regions of variation. This variation can occur anywhere in the genome, including areas that are not known to code for proteins. Over 99% of our DNA is noncoding DNA. Analyses of these noncoding regions has divulged repeated units of DNA that vary in length among individuals. STRs can be easily analyzed and compared between different individuals. The CODIS system uses 13 core STR loci that are now regularly used in the identification of individuals in the United States, and Interpol has identified 10 standard loci for the United Kingdom and Europe. Nine STR loci have also been identified for Indian populations.

As we learned earlier, an STR contains repeating units of a short (typically three- to four-nucleotide) DNA sequence. The number of repeats within an STR is referred to as an allele. For instance, the STR known as *D7S820*, found on chromosome 7, contains between 5 and 16 repeats of GATA. Therefore, there are 12 different alleles possible for the *D7S820* STR. An individual with *D7S820* alleles 10 and 15, for example, would have inherited a copy of *D7S820* with 10 GATA repeats from one parent, and a copy of *D7S820* with 15 GATA repeats from his or her other parent. To further simplify this, you have inherited *D7S820* with 15 GATA from your mother and *D7S820* with 10 GATA from your father. In this case, you would be heterozygous for *D7S820*. If you recall from Chapter 2 independent assortment during meiosis, then you can imagine each of your siblings may have different combinations of this particular STR. In this example, since there are 12 different alleles for this STR, there will be 78 different possible genotypes or pairs of alleles. Specifically, there will be 12 homozygotes, in which the same allele will be received from each parent, as well as 66 heterozygotes, in which the two alleles will differ.

Applied Biosystems GlobeFiler™ PCR Amplification kit uses 21 autosomal STR loci, meaning, all these are located on nonsex somatic diploid DNA chromosomes. It also contains 1 STR specific for the Y chromosome (*DYS391*). This is done to make sure that in case of mutation in the amelogenin locus, the forensic lab does not make a mistake in identifying the sex of the unknown DNA. As we learned in a previous chapter, sex identification is a crucial aspect of DNA genotyping or profiling. This kit also has the amelogenin STR as well as an additional STR to identify any mutation in the Y chromosome. These 24 loci also include all the 13 original STR markers so the old database does not become invalid.

In Chapter 7 we will learn how the whole process of STR DNA analyses is carried out and how we interpret the results of an **electropherogram**.

6.2 Learning Cycle 1

Can you use the Punnett square (**Figure 6.2**) and draw on a piece of paper all 78 genotypes and circle all the homozygous alleles?

Once you finish doing this exercise, you will appreciate the next topic, the statistical analyses of genotyping.

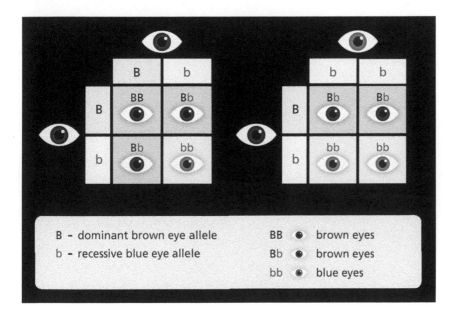

Figure 6.2   Punnett square.

6.3 What Is a Punnett Square?

A *Punnett square* is a visual representation of a cross. The genotypes of the parents are denoted along the top and the side of the grid. The possible genotypes of the offspring are obtained by combining the different alleles in the grid.

In **Figure 6.2**, the Punnett square shows an example of a cross between a heterozygous father and a homozygous dominant mother.

Chapter 7

How STRs Are Analyzed in a Forensic Laboratory

Workflow:

Extract DNA

Quantify DNA

"DNA quantification"

Perform PCR

"Prepare the amplification kit reactions"

"Perform PCR"

Run Capillary Electrophoresis

"Set up the instruments for electrophoresis"

"Prepare samples for electrophoresis"

Analyze Data

"Set up the GeneMapper™ ID-X Software for analysis"

▼

"Create an analysis method"
"Create a size standard definition file if needed"

▼

"Analyze and edit sample files with GeneMapper™ ID-X Software"
"Examine the electropherogram"

7.1 DNA Extraction and Analysis

To perform a forensic DNA analysis, DNA is first extracted from a sample. Just 1 ng of DNA is usually enough to obtain good STR data. The region containing each STR is then amplified by PCR to generate billions of identical copies of each of the 24 STR sequences shown later in **Table 7.1**. After PCR, CE is carried out on the amplified STR DNA sequences. The CE migrates the DNA sequences according to the size of the DNA fragments, giving an overall profile of all 24 STR sizes (alleles). The 24 core STRs vary in length from 100 to 300 bases, allowing even partially degraded DNA samples to be successfully amplified and then analyzed.

The costs of analysis, both in time and reagents, are significantly reduced by the amplification of all 24 STRs in a multiplex PCR reaction. The GlobeFiler™ kit uses six

Table 7.1 GlobeFiler™ and GlobeFiler™ IQC Kit Loci and Alleles

Locus Designation	Chromosome Location	Alleles Included in Allelic Ladder	Dye Label	Repeat
D3S1358	3p21.31	9, 10, 11, 12, 13, 14, 15, 16, 17, 18, 19, 20	6-FAM™	AGAT
vWA	12p13.31	11,12, 13, 14, 15, 16, 17, 18, 19, 20, 21, 22, 23, 24		TCTA/TCTG
D16S539	16q24.1	5, 8, 9, 10, 11, 12,13, 14, 15		GATA
CSF1PO	5q33.3–34	6, 7, 8, 9, 10, 11, 12, 13, 14, 15		AAAG
TPOX	2p23–2per	5, 6, 7, 8, 9, 10, 11, 12, 13, 14, 15		AATG
Y indel	Yq11.221	1, 2	VIC™	
Amelogenin	X: p22.1–22.3 Y: p11.2	X, Y		X, Y
D8S1179	8q24.13	5, 6, 7, 8, 9 10, 11, 12, 13, 14, 15, 16, 17, 18, 19		TCTA/TCTG
D21S11	21q11.2-q21	24, 24.2, 25, 26, 27, 28, 28.2, 29, 29.2, 30, 30.2, 31, 31.2, 32, 32.2, 33, 33.2, 34, 34.2, 35, 35.2, 36, 37, 38		TCTA/ TCTG
D18S51	18q21.33	7, 9, 10, 10.2, 11, 12, 13, 13.2, 14, 14.2, 15, 16, 17, 18, 19, 20, 21, 22, 23, 24, 25, 26, 27		AGAA
DYS391	Yq11.21	7, 8, 9, 10, 11, 12, 13		ACTA

(Continued)

Table 7.1 (Continued)

Locus Designation	Chromosome Location	Alleles Included in Allelic Ladder	Dye Label	Repeat
D2S441	2p14	8, 9, 10, 11, 11.3, 12, 13, 14, 15, 16, 17	NED™	TCTA/ TCAA
D19S433	19q12	6, 7, 8, 9, 10, 11, 12, 12.2, 13, 13.2, 14, 14.2, 15, 15.2, 16, 16.2, 17, 17.2, 18.2, 19.2		AAGG/ TAGG
TH01	11p15.5	4, 5, 6, 7, 8, 9, 9.3, 10, 11, 13.3		TCAT
FGA	4q28	13, 14, 15, 16, 17, 18, 19, 20, 21, 22, 23, 24, 25, 26, 26.2, 27, 28, 29, 30, 30.2, 31.2, 32.2, 33.2, 42.2, 43.2, 44.2, 45.2, 46.2, 47.2, 48.2, 50.2, 51.2		CTTT/ TTCC
D22S1045	22q12.3	8, 9, 10, 11, 12, 13, 14, 15, 16, 17, 18, 19	TAZ™	ATT
D5S818	5q21–31	7, 8, 9, 10, 11, 12, 13, 14, 15, 16, 17, 18		AGAT
D13S317	13q22–31	5, 6, 7, 8, 9, 10, 11, 12, 13, 14, 15, 16		TATC
D7S820	7q11.21–22	6, 7, 8, 9, 10, 11, 12, 13, 14, 15		GATA

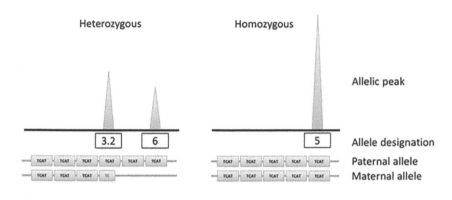

Figure 7.1 Homozygous vs. heterozygous allelic peaks.

different colors of TaqMan probes. When an STR migrates, a camera records the size and color of the STR and projects it on an electropherogram. If the DNA sample is homozygous for that STR, there will be ONLY ONE peak in the electropherogram. However, if the DNA sample is heterozygous for that locus, there will be two peaks. The following figure illustrates this concept.

In **Figure 7.1**, the DNA of the individual to the left possesses heterozygous peaks. Six *TCAT* STR were inherited from the father, and three from the mother. Whereas on the right side of the figure only one peak is shown, recording an individual who is heterozygous for *TCAT* with five *TCAT* repeats inherited from both parents.

In **Figure 7.2**, the electropherogram (in blue) of DNA from an unknown donor is homozygous for *D21S1435*, with a DNA sequence length of 175 bp, whereas for *D21S11* there are two peaks, 218 bp and 222 bp.

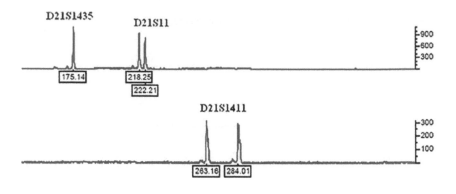

Figure 7.2   Electropherogram of unknown samples compared to a known sample.

The bottom electropherogram (in red) illustrates that the donor's DNA is heterozygous for *D21S1411*, with two peaks of 263 bp and 284 bp length.

Depending on the complexity of the repeat unit, the different alleles of an STR can vary by as little as a single nucleotide. For example, the *D7S820* STR, mentioned earlier, is relatively simple and contains between 5 and 16 repeats of GATA. Another STR, *D21S11*, has a more complex repeat pattern consisting of a mixture of tetra- and trinucleotides, and it, therefore, has alleles that vary in size by a single base pair. Because of the need to differentiate single-base differences from PCR products, highly sophisticated automated DNA sequencing technologies and software are used to recognize allele patterns by comparing them to a known "ladder" or "standards."

7.2 Making an STR Match

To match, for example, crime scene evidence to a suspect, a lab would determine the allele profile of the 24 core STRs for both the evidence sample and the suspect's sample. If the STR alleles do not match between the two samples, the individual would be excluded as the source of the crime scene evidence. However, if the two samples have matching alleles at all 24 STRs, a statistical calculation would be made to determine the frequency with which this genotype is observed in the population. Such a probability calculation considers the frequency with which each STR allele occurs in the individual's ethnic group. Given the population frequency of each STR allele, a simple Hardy-Weinberg calculation gives the frequency of the observed genotype for each STR. Multiplying together the frequencies of the individual STR genotypes then gives the overall profile frequency. If significant STRs are matched but due to degraded DNA some are lost, the probability that any-one else could have that specific DNA profile is extremely unlikely. The statistical probability would be 1 in 50 trillion!

In **Figure 7.3** we present a GlobeFiler™ profile obtained from a DNA sample. The GlobeFiler™ system uses six different colors and resolves each of the STR by the size of the amplified DNA. In the DNA profile depicted next, you will see all 24 STRs, resolved according to size range, and the six different-colored probes. The crime scene DNA profile can be compared to a suspect's DNA.

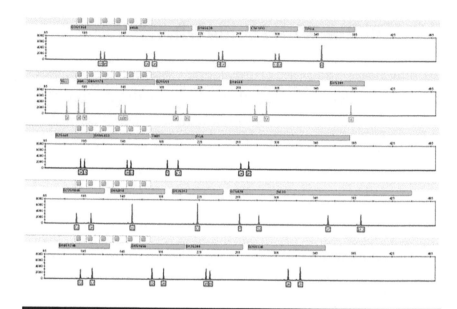

Figure 7.3   Example of a DNA profile obtained from GlobeFiler™ profile kit.

7.3 What Are We Amplifying?

The GlobeFiler is designed to amplify various simple STR (i.e., *AGAT*) vs. complex STR (i.e., TCTA/TCTG).

7.4 Confounding Circumstances of DNA Profiling

Sometimes, DNA from crime scene evidence is in a very small quantity, poorly preserved, or highly degraded, so only a partial DNA profile can be obtained. When fewer than 21 autosomal STR loci are examined, the overall genotype frequency is higher, therefore making the probability of a random match higher as well. For instance, if data are only obtained for the first four STRs listed in **Table 7.1** the likelihood of encountering this genotype would be roughly 1 in 331. In this instance, prosecutors would need additional types of evidence against a suspect to convince a jury that he or she was the source of the evidence sample. In addition, if an individual happens to have STR alleles that are very common in his or her ethnic group, the genotype frequency can also be higher, even when all the core 21 STR loci are examined. It is also important to note that crime scene samples sometimes contain DNA from several different sources. This can result in courtroom DNA evidence being less convincing.

The CODIS software contains multiple different databases that allow searches for various information, such as missing persons, convicted offenders, a convicted person on death row, and forensic samples collected from crime scenes of unknown origins. Each state, and the federal system, has different laws for collection, sharing, and analysis of information contained within their database. However, for privacy reasons, the CODIS database does not contain any personal identifying information, such as the name associated with a DNA profile. The uploading agency is notified of any hits to their samples and is then asked to provide additional information on the "hit" DNA personal information pursuant to their laws.

7.5 The Statistical Strength of a 24-STR Profile

Within the United States, the 21-STR profile is widely used for identification, and this DNA-based profiling technology is now routinely employed to identify the source of DNA at a crime scene, to identify human remains, to establish or exclude paternity, or to match a suspect to a crime scene sample. In the next chapter, we will elaborate on the expansion of STR markers and learn why this has become necessary.

To utilize STR information as a means of human identification, the FBI has established the frequency with which each allele of each of the 21 core STRs naturally occurs in people from different ethnic backgrounds. To this end, the FBI analyzed DNA samples from hundreds of unrelated Caucasians, African Americans, Hispanics, and Asian individuals. Assuming that all 21 STRs follow the principle of independent assortment during meiosis and that the population randomly mates, a statistical calculation based upon the FBI-determined STR allele frequencies reveals that the probability of two unrelated Caucasians having identical STR profiles, or so-called

"DNA fingerprints," or DNA genotypes, is approximately 1 in 575 trillion.

7.6 The Complication of Sex Determination

In a previous chapter, we considered DNA sex determination from samples derived from a crime scene, such as from the buried remains of an unidentified body found at a mass casualty site. We discussed the use of the amelogenin sex STR to determine the sex of the source DNA. However, more recently, it has been discovered that the amelogenin gene (*AMELY*) acquires lots of mutations, and in some humans, it is deleted altogether. Therefore, as you might have noticed already, we use three DNA markers instead of just one DNA marker to identify the male sex. These are Y STR locus, Y indel locus, and sex-determining marker (amelogenin).

7.7 Can a Nonhuman DNA Contamination Mismatch an STR?

The GlobeFiler™ as well as other STR primers are selected to only amplify "human" DNA. Therefore, the DNA profiling kits are designed in a way that they exclude amplifications of gorillas, chimpanzees, macaques, mice, dog, sheep, pigs, rabbit, cat, horse, hamster, rat, chicken, and cow. These kits also exclude amplifications of any bacterial DNA, since bacterial contamination is not unusual in a DNA sample. These include DNA from *Candida albicans, Staphylococcus aureus, Escherichia coli, Neisseria gonorrhoeae, Bacillus subtilis*, and *Lactobacillus rhamnosus* (**Figure 7.4**).

Of note, the *AMEL* gene is partially conserved among the large primates (i.e., gorilla, chimpanzee, macaque), and DNA contaminations from these species may amplify *AMEL*. However, the STR sizes from X and Y from the large primates are much larger in length and can easily be excluded. This is shown in the figure that follows.

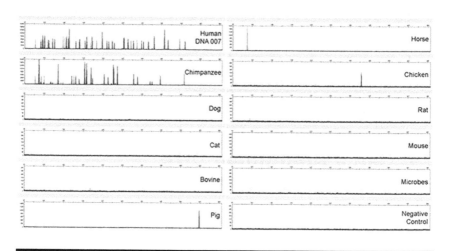

Figure 7.4 STR profiles of various animals compared with human DNA.

Chapter 8

Complexities of Sex Determination

Since the advent of STR kits, forensic scientists have been using amelogenin as the major tool to determine the sex of donor DNA. For sex determination, we amplify a precise fragment of DNA on X and Y chromosomes and then measure the sizes of each fragment to determine sex (**Figure 8.1**). Depending on the STR kit used, the sizes of X and Y will vary. By using GlobeFiler™, *AMELX*'s will be six-base-pair smaller than *AMELY.* From a male DNA, we will recover "two bands" of DNA, at 106 bps for X and 112 bps for Y. From female DNA (*AMELX*), only one 106 bp fragments will show on the electropherogram, as illustrated in **Figure 8.2**.

DOI: 10.1201/9781003182498-9

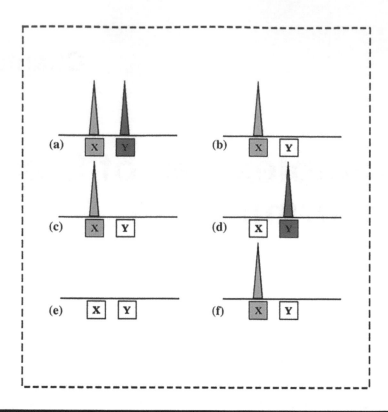

Figure 8.1 Probable results and problems associated with amelogenin typing in normal as well as compromised samples: (a) normal male, (b) normal female, (c) male with Y deletion, (d) male with X deletion, (e) female with X deletion in both the chromosomes, (f) female with X deletion in any one chromosome.

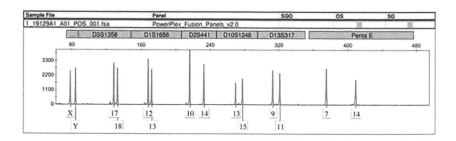

Figure 8.2 The electropherogram shows X and Y peaks, indicating the source of DNA is from a male. DNA from a female source will give a single peak. Since a female carries two X chromosomes, the peak will be higher, as shown in the next diagram.

8.1 Complications in Determining the Sex of a Donor DNA

The amelogenin marker has been used for a long time; however, its efficiency came into question when reports of false detection of males as females began to appear in various scientific journals and during forensic science conferences. One of the main reasons for this was the occurrence of mutations and deletions in the areas of *AMELY* where the PCR primers bind to amplify the DNA gene segment, resulting in false negative results. Therefore, the male's *AMELY* DNA gene segment carrying deletions and mutations would be identified as females. Inconsistencies in the amelogenin-based sex determination due to mutation or deletion in *AMELY* can result in different false-positive and false-negative results. Unfortunately, recent data show that mutations and deletions in the amelogenin region of the Y chromosome are more common than in the X chromosome. These deletions and mutations are common in groups that marry within their own specific ethnic or caste system. In most cases, these kinds of ethnic/caste unions, known as *endogamy*, prohibit marriage outside of their own group. Failure to determine biological sex can range up to 10% in an *endogamous* population. Therefore, additional genotyping kits must be used. GlobeFiler™, which we discussed previously, among a few other kits, uses different sets of primers and an additional two biomarkers to determine the sex of the donor DNA (**Figure 8.3**).

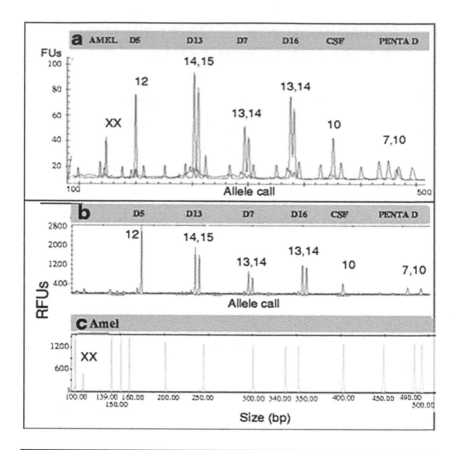

Figure 8.3 Electropherogram from a female DNA donor, showing a single AMELX peak.

8.2 Y Deletion and Endogamy

Endogamy is common in many cultures and ethnic groups. Several religious and ethnic groups are traditionally more endogamous. Endogamy, as distinct from consanguinity (where cousins marry each other), may result in the transmission of genetic disorders, the so-called founder effect, within the relatively closed community.

Endogamy can serve as a form of self-preservation or resistance against integrating and completely merging with surrounding populations. Minorities can use it to stay ethnically homogeneous over a long time as distinct communities within societies that have other practices and beliefs. There are numerous examples of endogamy and may include Amish, Assyrians, Orthodox Jews, Mormons, Parsis, Yazidis, and small Muslim sects of Ismailis, Bohra, Rajputs of India, and of course, the religiously mandated caste system that segregates different groups by religious practices of Hinduism.

The isolationist practices of endogamy may lead to a group's extinction, as genetic diseases may develop that can affect an increasing percentage of the population. However, this disease effect would tend to be small unless there is a high degree of close inbreeding or if the endogamous population becomes very small in size (e.g., only ~7,000 Yazidis are left).

Most of the cases of amelogenin-based sex determination failure result from the absence of *AMELY* in the genotyping results when the subject had the Y chromosome. These discrepancies are due to the deletion of nucleotide sequences on the Y chromosome that include the *AMELY* region. In cases of Y deletion, known male samples show only the presence of *AMELX* with 6 bp deletion, resulting in the appearance of only one peak after genotyping. In these cases, the presence of the Y chromosome can be assured either by karyotyping or using alternative markers present on the Y chromosome. Many studies involving Indian populations have confirmed *AMELY* deletion at a higher rate in isolated groups. Y deletion is present in almost all populations throughout the world. As a major part of the Y chromosome does not undergo recombination, which we have learned in Chapter 2, any deletions or mutations in the Y chromosome are passed onto subsequent generations, resulting in the formation of geographically specific haplogroups. The haplogroups belonging to the Indian and Asian ancestry have been found to have a higher frequency of Y deletion.

8.3 X Deletion

Since females have two X chromosomes, *AMELX* is a rare phenomenon as compared to *AMELY* mutations and deletions. Mutation in *AMELX* can be the cause of *AMELX* absence in males, but it would not affect the identification of a male since there would be normal amplification of *AMELY*. Additionally, in the case of females, *AMELX* absence is difficult to detect as it is extremely rare to have a mutation in both the homologous chromosomes, and in cases of mutation in a single chromosome, a single peak would be observed like a normal female, as shown in the following figure. Nevertheless, rarely, forensic scientists have found a complete absence of *AMELY* loci.

Most of the currently used human STR kits such as InnoTyper21™ kit, AmpFLSTR™ Identifiler™ kit, and PowerPlex® 21 system do not possess any alternative sex-determining markers. However, a few new generation kits such as GlobalFiler™ (Thermo Scientific), PowerPlex® Fusion 6C (Promega Corp.), Investigator 24plex GO! Kit (Qiagen), and SureID® PanGlobal (Health Gene Tech.) contain *DYS391, DYS570, DYS576*, and Y InDel as alternative sex-determining markers. However, when *AMELY* deletion occurs, simultaneous deletion of *DYS570* and *DYS576* occurs as they are present within the *AMELY* region on the Y chromosome. Additionally, the position of *DYS391* is very close to the *AMELY* region, which increases the chance of codeletion of *DYS391* along with other Y-STR markers.

As we have discussed in a previous chapter, that GlobeFiler™ contains primers for 21 autosomal STR loci (i.e., *D3S1358, vWA, D16S539, CSF1PO, TPOX, D8S1179, D21S11, D18S51, D2S441, D19S433, TH01, FGA, D22S1045, D5S818, D13S317, D7S820, SE33, D10S1248, D1S1656, D12S391, D2S1338*), for one Y-STR (*DYS391*), for one insertion/deletion polymorphic marker on the Y chromosome (Y indel), and for amelogenin (sex-determining marker).

While the amelogenin sex test is accurate in general, other Y chromosome markers, such as *SRY,* are being used for less-ambiguous gender identification. The *SKY* gene is also known as the "testis-determining factor" (TDF). The *SKY* gene is responsible for initiating male sex determination in all placental and marsupial mammals. Therefore, if not certain, one can use *SKY* as another marker to confirm male origin of a DNA sample in question.

Chapter 9

Y-STR and Paternal Ancestry

In 1998, the scientific journal *Nature* published the results of DNA tests designed to shed new light on questions first asked some 200 years earlier: Did Thomas Jefferson have a relationship with a woman who was his slave? Did that relationship produce children?

Y-STRs are relatively new DNA markers and often used in forensics, paternity, and genealogical DNA testing (**Figure 9.1**). Using Thomas Jefferson's DNA and the DNA from the male descendants of Sally Hemings, it was discovered that the famous president fathered at least one male child from his slave Sally Hemings, and most likely all her six children. Y-STR loci are embedded in the male Y chromosome. Compared to the 21 autosomal STRs we described previously, these Y-STRs provide a weaker analysis since the Y chromosome is

DOI: 10.1201/9781003182498-10

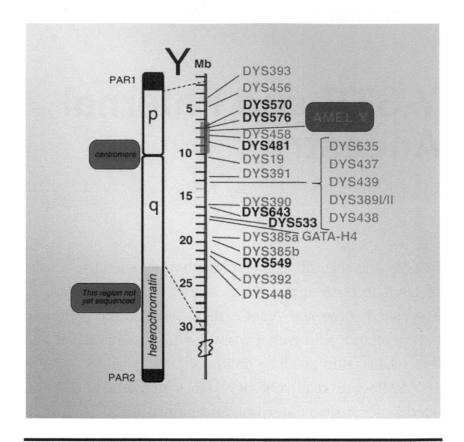

Figure 9.1   Illustration of currently known Y-STRs.

Source: "Y STR Positions along Chromosome," nist.gov.

only found in males. Also, the Y chromosome in any paternal line is virtually identical. Autosomal STRs provide a much stronger analytical power because of the random segregation and the crossover between pairs of chromosomes during the meiosis process.

The Y chromosomes are similar in all males from a family tree, meaning, that your Y chromosome is virtually identical to your great-great-grandfather's Y chromosome. In genetic genealogy, the use of YSTRs makes a fantastic tool. Therefore, the Y-STR database is a wonderful way to locate one's family tree. The Family Tree DNA database carried many genetic databases on Y-STR (**Figure 9.2**).

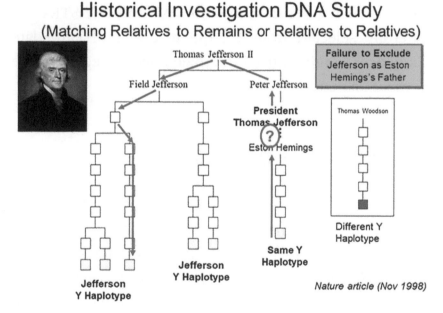

Figure 9.2   Illustration of Thomas Jefferson's Y chromosome and similarity to the Y chromosome from two of the "descendant" male children born from Sally Hemings.

9.1 Y-STR Markers

Currently, there are only a few YSTR profiler kits on the market. The first multiplex PCR kit for routine application became available with 12 STR loci. In the following years, the number of markers increased steadily. Currently, different kits and supplementary multiplexes can be combined to amplify over 40 Y-STR sequences, a required number to pursue a strategy "typing until exclusion" in forensic casework. All these Y-STRs are carefully selected from a large reservoir of candidate sequences obtained from the male-specific region of the Y chromosome.

9.2 Reference Databases

One of the most important events with Y-STR was the establishment of online accessible population databases which consist of Y-STR profiles typed in reference populations (**Figure 9.2**). The largest known database, the YHRD, now includes more than a hundred national databases which were populated in the past 20 years and continue to grow at a steady pace. The YHRD project started in 1999 with a collection of 3,825 haplotypes typed in 48 populations for up to 13 Y-STRs. Just to remind you, when we state "a haplotype," it is defined as a group of Y chromosome alleles inherited from one's father. Currently, the

59th edition of the YHRD is online, with 265,324 profiles from 1,251 populations typed for up to 29 Y-STRs and many single nucleotide polymorphisms (SNPs), meaning, their sequence differs only at a single bp. Resources include primary data, such as regularly and irregularly spaced, null and multiple alleles, haplotypes in different formats, mutation rates, and Y-SNP data. These data facilitate statistical analyses for a range of subjects, including kinship and mixture analysis and ancestry prediction based on Y haplogroups.

Of note, the statistical methods relying on the independence of loci, as we observe in the 22 autosomal chromosomes, cannot be applied to Y chromosomal profiles. All Y-STRs, which constitute a haplotype, are linked together and are not inherited independently but all together along paternal lineages. In a criminal case, it is, therefore, safe to assume that besides the true trace DNA donor, other men with the identical profile exist, including close relatives, and are expected to carry identical haplotypes, but also, unrelated males may display the same haplotype as the true donor.

Currently, the 17-locus YFiler® haplotype is the most used YSTR kit. It is important to know that if a YFiler profile matches with a suspect in Western Europe, the probability of occurrence of this haplotype would be around 1 in 50,000 (a population of ≈400 million). According to this number, roughly 8,000 men unrelated to the suspect could carry an exact identical haplotype in Western Europe. The same haplotype has a chance to occur in the Chinese Han population with ≈1 in 236 million, or 5 men in a population of ≈1.2 billion. Of note is that Y-STR testing passed admissibility standards, for example, in the USA, because its analysis and statistical treatment are no different to autosomal STRs. Typing, nomenclature, mutation mechanisms are similar, and even the dependency structure of Y-STRs is not so different; the Y chromosome is one forensic marker (though extremely variable) among others. The shared identity by parentage makes the interpretation of haplotypes difficult, but as in the autosomal case, more STRs, good databases, and statistical probability methods can provide reliable and fair results.

9.3 Ancestry Inference Using the Y Chromosome

The most wonderful use of YSTR is the ability to predict male ancestry in a forensic context based on the universal phylogenetic tree of Y chromosomes. Recently, a very extralarge phylogeny tree, incorporating 13,261 high-confidence SNPs, retrieved from 3.7 Mb of the male-specific part of the Y chromosome, sequenced in 448 human males, has been published and can be used to trace the genealogies of people.

9.4 Investigative Leads Provided by Y Chromosome Testing

The most used Y chromosome analyses are being conducted in cases where rape crimes are committed where a male has left little to no sperm and female vaginal epithelial DNA must be differentially separated from a very small amount of male DNA. Here, the Y-STRs and Y-biomarkers can detect hidden male DNA on a strong background of female DNA. A Y-STR screening is, therefore, carried out in all cases where male DNA, left by a perpetrator, is expected but not found in regular autosomal STR screening. Samples with full Y could then be sensitively reanalyzed for autosomal markers to develop a profile eligible to search in a national DNA database. Most databases do not upload Y-STR profiles because of the ambiguity of a match, which could also match a group of paternal relatives. In severe crime cases, Y chromosome analysis can be seen as a "last-resort" technique to collect reliable information on an unknown perpetrator and to narrow down the number of suspects.

9.5 Summary

Currently, Y chromosome markers are used to assist forensic investigations in a broad range of casework, either for rapid sexing of traces, for the identification and profiling of miniscule amounts of male DNA in mixtures with very large amounts of female DNA, for tracing male lineages in relationship testing, or for analyzing the deep-rooting ancestry of unidentified corpses or unknown trace donors. Of course, in genealogy, an example of which we showed at the beginning of this chapter—the Thomas Jefferson and Sally Hemings children case resolved by using Y-STR. Like any forensic DNA analysis, judging the significance of a match depends upon the population frequency of the profile. As has been discussed in this chapter, Y chromosomes are highly geographically limited, and there is the further complication that all members of a patrilineage are expected to share Y haplotypes. The forensic community has been attempting to establish large quality-assured and publicly accessible databases of Y haplotypes to solve several issues with YSTR.

Chapter 10

Mitochondrial STRs

As we learned in the previous chapter, the Y chromosome comes only from the father, and it is a great tool to find one's paternal ancestry. But we received all our mitochondria ONLY from our mothers! This means that we can trace our maternal ancestors by using mitochondrial DNA.

First things first—mitochondria are found in the cytoplasm of all living cells. They are not chromosomes, but we inherit them from our mothers. They can replicate independently, and unlike chromosomes, one can find hundreds of mitochondria in a single cell. Their number changes according to cellular energy needs. They are responsible for most of the chemical reactions that place in respiration, and respiration releases energy that the cell can use. They generate most of our energy in the form of ATP. The cells that require a lot of energy have many mitochondria compared to cells that are metabolically silent. Evolutionarily, the mitochondria originated

DOI: 10.1201/9781003182498-11

as bacteria that developed symbiotic life with eukaryotic organisms over a billion years ago (**Figure 10.1**).

Structurally, mitochondria are small cytoplasmic organelles (miniature organs within the cell), measuring between 0.75 and 3 micrometers, and are not visible under the microscope unless they are stained. Unlike other organelles, they have two membranes, an outer one and an inner one. Each membrane has different functions. Mitochondria are split into different compartments or regions, each of which carries out distinct roles.

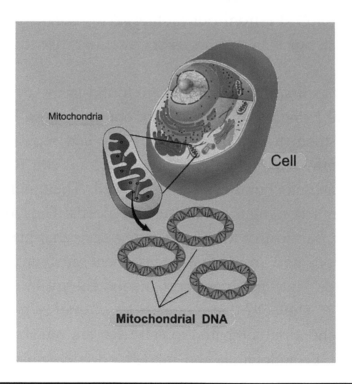

Figure 10.1   Mitochondrial DNA.

Some of the major regions and functions include:

Outer membrane. Small molecules can pass freely through the outer membrane. This outer portion includes proteins called porins, which form channels that allow proteins to cross. The outer membrane also contains several enzymes with numerous functions.

Intermembrane space. This is the area between the inner and outer membranes.

Inner membrane. This membrane has numerous proteins embedded into it with various functions. Because there are no porins in the inner membrane, it is impermeable to most molecules. Molecules can only cross the inner membrane in special membrane transporters. The inner membrane is where most ATP is created, and it is the reason we call mitochondria the powerhouse of life.

Cristae. These are the folds of the inner membrane. They increase the surface area of the membrane, therefore increasing the space available for chemical reactions to create energy.

Matrix. This is the space within the inner membrane, containing hundreds of enzymes which play a key role in the production of ATP. Mitochondrial DNA (mtDNA) is housed in the innermost area of the matrix (**Figure 10.2**).

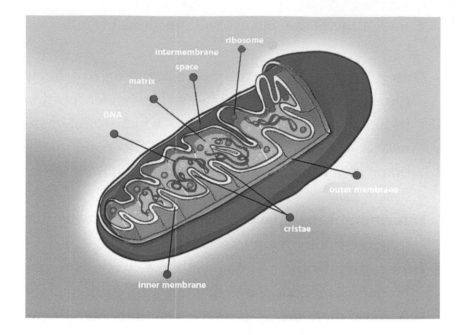

Figure 10.2   An illustration of mitochondria and the location of its DNA.

Different cell types have different numbers of mitochondria. For instance, mature red blood cells have no mitochondria, whereas liver cells can have more than 2,000. Cells with a high demand for energy tend to have greater numbers of mitochondria. Around 40% of the cytoplasm in heart muscle cells contain mitochondria. Although mitochondria are often drawn as oval-shaped organelles, they are constantly dividing (fission) and bonding together (fusion). So these organelles are linked together in ever-changing networks. Also, in sperm cells, the mitochondria are spiraled in the midpiece and provide energy for tail motion. The sperm that has the highest compacted energy has a higher chance of reaching the egg. After one successful sperm reaches the egg, the head of the sperm penetrates the egg and within milliseconds this egg no longer accepts any other sperm. This phenomenon is illustrated on page 142.

10.1 Mitochondrial DNA

Mitochondrial DNA (mtDNA) is stored in the matrix. Unlike nuclear DNA, which contains around 3 billion bp, mtDNA consists of 16,569 bp. We receive our mtDNA only from our maternal side, and due to this, we can trace our maternal origin to Africa around 200,000 years ago. Our human mother referred to as "The Mitochondrial Eve" resided in Africa, and we are her descendants.

The real reason that we inherit mitochondria is that when a sperm enters an egg, it leaves the midpiece behind, where all the mitochondria are located, as shown in **Figure 10.3**.

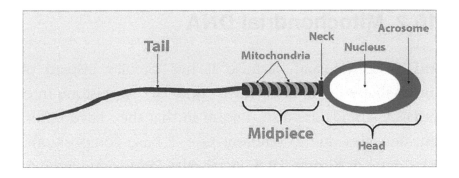

Figure 10.3 **An illustration of sperm with locations of its mitochondria and haploid DNA in the sperm head.**

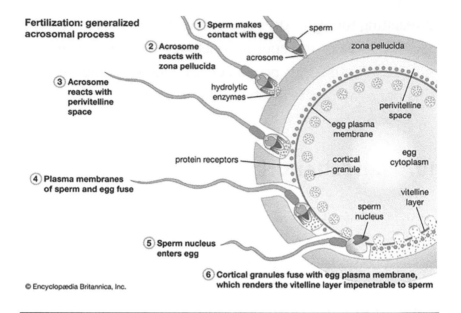

Figure 10.4 An illustration of fertilization.

So it is fair to say that a sperm breaks its neck to get inside the egg (**Figure 10.4**).

10.2 Mitochondrial DNA

MtDNA is unusual because it has circular instead of linear-shaped chromosomes, and it is histone-free. mtDNA strands are also unique in that they have different densities due to different G + T base compositions. As shown in **Figure 10.5**, its circular DNA is divided into two strands: heavy (H) and light (L) rings. The H strand encodes more information, with genes for two rRNAs

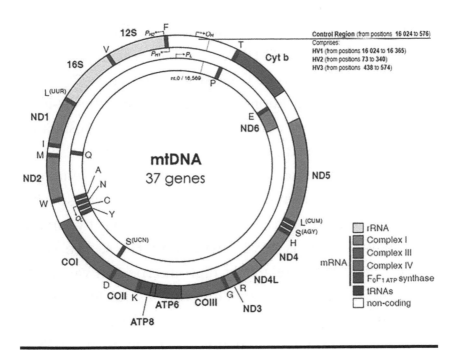

Figure 10.5 The human mitochondrial DNA genome with genes and control regions labeled.

Source: Adapted from Picard, Wallace, and Burelle (2016).

(12S and 16S), 12 polypeptides and 14 tRNAs, while the light (L) strand encodes 8 tRNAs and 1 polypeptide. All 13 protein products are part of enzyme complexes that constitute the oxidative phosphorylation system, or electron transport chain, that creates energy. Also, the mtDNA are intronless, meaning, they don't have any introns—segments of DNA that do not code for proteins and interrupt gene sequences; mtDNA consists of exons, or coding genes, and a very small region which is noncoding and known as a D-loop. The mitochondrial D-loop is a triple-stranded region found in the major noncoding region of mitochondrial DNA and is formed by the incorporation of a third short DNA strand known as 7S DNA. Mitochondrial replication is initiated from the D-loop region; mtDNA does not have any repair mechanism and, because of this, expresses a high mutation rate when compared to the nuclear genome. Many of the sequence variations between individuals are found in two specific segments of the control region, known as the hypervariable regions: that is, HV1, HIV2, and HIV3. The small size of mtDNA and relatively high interperson variability of the hypervariable regions are very useful features for ancestry tracing and other forensic testing purposes.

Each mitochondrion contains 2–10 copies of mtDNA, and each somatic cell can have up to 1,000 mitochondria; therefore, when the amount of the nuclear DNA is degraded or too small to carry out STR analysis, mtDNA typing can be used to look for polymorphic regions that are found in nuclear DNA. As mentioned earlier, the mtDNA is only maternally inherited; therefore, except for minor mutations, mtDNA sequences of siblings and all maternal relatives would be identical. This specific characteristic is very useful in forensic cases, for the analysis of the remains of a missing person, where known maternal relatives can provide some reference samples for a direct comparison to the mtDNA type. Due to the lack of recombination, mtDNA analyses can be used to trace the maternal relatives from several generations apart from the source of evidence and can also be used for reference samples.

The haploid and monoclonal nature of the mtDNA in most individuals simplifies the process of interpretation of the DNA sequencing results.

MtDNA sequences can be used to trace the ancestries of various ethnic groups. For example, most individuals from African populations, and especially those from sub-Saharan Africa, are categorized into one of the main haplogroup lineages that diverged from the macro-haplogroup L: L0, L1, L2, L3, L4, L5, and L6. The most observed haplogroups are L2a, L1c, L1b, and L3b. Over 90% of the individuals of the European and US Caucasian populations are categorized into ten main haplogroup lineages: H, I, J, K, M, T, U, V, W, and X. The main haplogroups found in individuals from Asian populations are haplogroups M and N.

10.3 Mitochondrial DNA Population Data and Databases

The most important mtDNA haplotypes database is the EDNAP Mitochondrial DNA Population Database (EMPOP, empop.online). EMPOP not only serves as a reference population database but also provides the most comprehensive resource, especially from the standpoint of the populations that are represented. When using EMPOP, the tool haplogroup browser represents all the established PhyloTree haplogroups in convenient searchable format and provides the number of EMPOP sequences assigned to the respective haplogroups by estimating mitochondrial DNA haplogroups using the maximum likelihood approach EMMA. PhyloTree provides an updated, comprehensive phylogeny of global human mtDNA variation, based on both coding and control region mutations. The complete mtDNA phylogenetic tree includes previously published as well as newly identified haplogroups; it is continuously and regularly updated and is available online at www.phylotree.org. At EMPOP, the geographical haplogroup patterns are provided via maps to visualize and better understand their geographical distribution.

10.4 Mitochondrial Diseases

MtDNA analysis may also be useful in detecting inherited diseases associated with mutations in mtDNA. Even though relatively short, there are around 3,000 different proteins associated with mtDNA, but only about 13 of these are coded on the mtDNA. Most of the 3,000 types of proteins are involved in a variety of processes other than ATP production, such as porphyrin synthesis. Only about 3% of the code is for ATP production proteins. This means most of the genetic information coding for the protein makeup of mitochondria is in chromosomal

DNA and involved in processes other than ATP synthesis. This increases the chances that a mutation that will affect a mitochondrion will occur in chromosomal DNA, which is inherited in a Mendelian pattern. Another result is that a chromosomal mutation will affect a specific tissue due to its specific needs, whether those may be high energy requirements or a need for the catabolism or anabolism of a specific neurotransmitter or nucleic acid. Because several copies of the mitochondrial genome are carried by each mitochondrion (2–10 in humans), mitochondrial mutations can be inherited maternally through mutations present in mitochondria inside the oocyte before fertilization, or through mutations in the chromosomes.

Mitochondrial diseases, which range in severity from non-noticeable to severe or even fatal, are most commonly due to inherited rather than acquired mutations of mtDNA. A mitochondrial mutation can cause various diseases, depending on the severity of the problem in the mitochondria and the tissue the affected mitochondria are in. Conversely, several different mutations may present themselves as the same disease, making them very hard to diagnose and trace. Some diseases are observable at or even before birth (many causing death), while others do not surface until late adulthood. This is because the number of mutants versus wild-type mitochondria varies between cells and tissues and is continuously changing. Because cells have multiple mitochondria, different mitochondria in the same cell can have different variations of mtDNA. This condition is known as heteroplasmy. When a certain tissue reaches a certain ratio of diseased versus normal mitochondria, a disease will present itself. The ratio varies from person to person and tissue to tissue. Mitochondrial diseases are very numerous and different. Apart from diseases caused by abnormalities in mitochondrial DNA, many diseases are suspected to be associated in part by mitochondrial dysfunctions, such as diabetes mellitus, forms of cancer and cardiovascular disease, lactic acidosis, specific forms of myopathy, osteoporosis, Alzheimer's disease, Parkinson's disease, stroke, male infertility, and the aging process (**Figure 10.6**).

Figure 10.6   An illustration of some of the diseases associated with mitochondria.

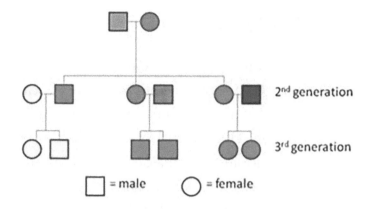

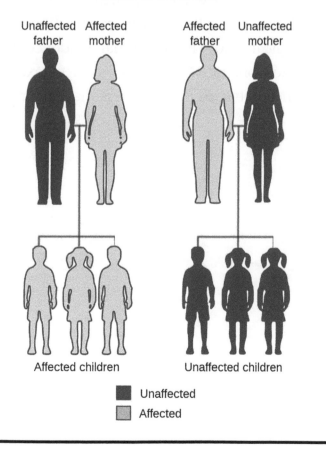

Figure 10.7 Mitochondrial inheritance.

10.5 Summary

MtDNA is essentially a matrilineage ancestral genetic marker (**Figure 10.7**). But the mtDNA genome can also provide information about ancestors, including health/disease information. Human mtDNA can also be used to help identify individuals. Forensic scientists occasionally use mtDNA comparison to identify human remains, and especially to identify older unidentified skeletal remains. Although, unlike nuclear DNA, mtDNA is not specific to one individual, it can be used in conjunction with other anthropological and circumstantial evidence to establish identification. mtDNA is also used to exclude possible matches between missing persons and unidentified remains.

Chapter 11

Forensic Serology

Whose Blood Is This? Is It Human?

11.1 What Is Forensic Serology?

During any forensic investigation usually, the very first thing an investigator must do is collect biological evidence. This evidence includes bloodstains, fresh or clotted blood, and trace amounts of blood on clothes and on human body parts. In addition, saliva, sweat, semen, breast milk, or other bodily fluids are all parts of forensic investigations carried out by an expert forensic serologist. Once the evidence is collected, it is crucial for medicolegal reasons to protect and secure the evidence. This includes securing the evidence in a way that is safe from environmental effects, such as heat, drying, microbial degradation or contamination, and human interference (**Figure 11.1**).

Forensic serology and forensic DNA analysis are linked together. When crime-related evidence is submitted to a forensic laboratory, first they are coded so a forensic

DOI: 10.1201/9781003182498-12

Figure 11.1   Blood evidence is a crucial part of forensic science.

scientist analyzing the evidence does not know which evidence belongs to which person. Once evidence is given a code, the evidentiary samples are handled by the forensic serologist to determine an identifiable body fluid, which is subsequently analyzed by DNA analysis to accurately link a fluid to a specific individual.

The ability to detect and identify every stain and, at the same time, to confirm the real human nature of each stain is crucial both for medicolegal purposes and for court documentation. Carrying out elaborate, expensive, and time-consuming STR-DNA analyses before making sure that the actual "stains" are of human origin can result in serious errors. Blood is one of the most common types of evidence found at crime scenes. First, a crime scene investigator will carry out one of the presumptive tests for blood, but a presumptive test does not give much information about the human origin of the stain; therefore, additional confirmatory tests must be performed.

If a blood sample is submitted to a crime laboratory, one of the important questions that need to be answered is, "Is it human blood?" Often, several blood specimens can come from domestic or other animals, especially when the body is found in an open area.

At the crime scene, it is important to establish the type, origin, and other characteristics of the blood/bloodstain. Blood pattern analysis is a separate field of forensic science and is very well fictionalized in the HBO series *Dexter*.

The initial investigations of recovered fluid from a crime scene are to find out whether it is from a human. Once the human origin of the blood/bloodstain, saliva, semen, and other bodily fluids is confirmed, further analysis is done to establish other characteristics. The bloodstain needs to be "blood-typed" first. What group does this human blood belong to? ABO blood typing establishes its blood type (i.e., A, B, AB, O). Proteins, enzymes, and antigens present in the blood of the individual are also analyzed. Blood testing is a highly precise and delicate task, and a serologist makes sure that the blood collected is devoid of any environmental source outside of the stain areas. A careful collection is vital, since blood-typing antigens are also found in other mammals, bacteria, wood, soil, dust, and elsewhere. In addition, microorganisms in the environment may enzymatically alter blood group antigens. Microorganisms like *Clostridium* spp., *Bacillus* spp., and *Aspergillus niger*, as well as coffee beans, can alter or eliminate certain blood group substances.

At a crime scene, a presumptive blood test is carried out to determine if the blood has a human origin (this goes for all bodily fluids). All the presumptive tests must be confirmed by the serologist at the crime laboratory by more precise methods, and all the tests must be approved by the federal agencies of the country where the tests are being performed.

In the following section, we present various methods used in forensic serology to identify human blood and distinguish it from animal blood or blood contamination.

11.2 Presumptive Blood Tests

Kastle–Meyer Test. The most common presumptive test carried out at the crime scene (when necessary) is the Kastle–Meyer test, which uses a chemical called phenolphthalein. Phenolphthalein binds to hemoglobin and produces a pink color. It is only a presumptive test, and this chemical can bind other substances like saliva, pus, malt extract, vegetable extract, and certain metals. To conduct the test, the CSI uses a cotton swab and applies a dry or fresh blood sample to a wet cotton swab. The cotton swab is then dried with alcohol, and phenolphthalein is added, followed by hydrogen peroxide. If hemoglobin is present, the swab will turn pink (**Figure 11.2**).

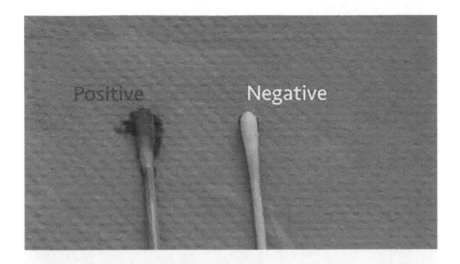

Figure 11.2 Full video demonstrating the procedure at the Kastle–Meyer blood test.

Source: YouTube.

11.3 Luminol Test

You might have seen in many of the CSI shows that a law enforcement officer or an investigator carries a UV light to detect blood at a crime scene. To detect invisible stains of blood, the CSI first sprays luminol on the areas where one suscepts the presence of invisible blood. The luminol binds with hematin and produces a luminescence color. Luminol-sprayed areas are best visualized in the dark. If blood is present, the positive areas will show a blue glow. This can be documented by photographs. Interestingly, the glow of an older bloodstain lasts much longer than that of fresh blood. The chemical also binds to certain

Figure 11.3   Luminol to detect bloodstains at the crime scene.

metals like iron and copper, and horseradish and bleach can give false positive results (**Figure 11.3**).

11.4 Confirmatory Tests for Blood, Saliva, Semen, and Other Body Fluids

11.4.1 *Confirmation of Human Blood*

11.4.1.1 *Rapid Stain Identification–Blood (RSID–Blood)*

When bloodstains are found at a crime scene using one of the presumptive tests described previously, then it is important to confirm if the blood is of human origin. There are several methods that forensic laboratories utilize. One of these is the **rapid stain identification–blood** (RISD–blood). This test is very similar to what you might have seen used to detect influenza A infection in a physician's office, where a health-care worker uses saliva or nasal secretion from a sick child or adult to determine whether they have flu. Several tests have been developed by the FBI to detect blood, semen, saliva, and other body fluids. These tests are based on specific antibody conjugation reactions to form colored complexes in the presence of specific biological proteins or antigens. To perform a test, the sample in question is mixed with a buffer specific to the RSID test. The sample is then deposited onto the sample window, where it is pulled through the test and control regions by a paper wick via passive diffusion. The RSID unit is embedded with fluid-specific antibodies conjugated to colored complexes that will redissolve and diffuse at this time. For example, to detect blood, the strip is embedded with antibodies to glycophorin A, a red blood cell membrane antigen. The blood RSID uses two antihuman glycophorin A (GPA) monoclonal antibodies. If the blood is of human origin, it will result in a positive reaction with glycophorin antigens inside the strip and develop a color confirming the human origin of the blood found at the crime scene. A positive control is always included; therefore, a color line will form at the control region regardless if

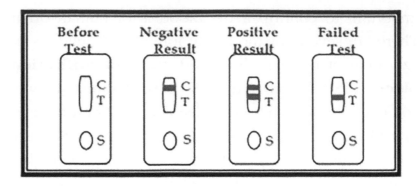

Figure 11.4 Rapid stain identification test.

Source: RSID test for human blood, Google Search.

the sample contains body fluid or not, as it acts as an internal control.

Positive and negative controls are also performed alongside each test to confirm the reliability of the test. Positive controls are swabs taken from known bodily fluids, while negative controls are buffer without any biological sample (**Figure 11.4**).

11.5 Christmas Tree Stain to Identify Human Sperm

For the visual identification of sperms, we use two main reagents, successively, which produce a distinctive stain. Picroindigocarmine stains the neck and tail portions of the sperm in green and blue, while nuclear fast red gives the sperm heads a red color and the tip of the heads a pink color. Of note, sperm cells deteriorate quickly after ejaculation. Sperm survival will depend on the surrounding environment and type of surface. The sperm tails are the most susceptible to damage and will break down first. Therefore, the analyst must be trained to make visual distinctions between sperm heads and other types of cells in the mix. Other cells will also stain red (**Figure 11.5**).

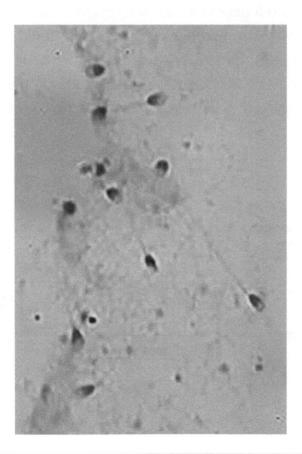

Figure 11.5 Christmas tree stain for human sperm.

Source: Christmas tree stain to identify human sperm, Google Search.

11.6 RSID Test for Semen

This rapid test is similar to the blood identification test, except this test uses an antigen called **seminal vesicle-specific antigen**, or **semonogelin**. This antigen is unique to human semen; therefore, a positive test is definitive, since it does not cross react with other bodily fluids in males and females or with semen from other mammals.

The CSI team member can use a handheld electronic reader to read the results of confirmatory tests for blood or semen. The reader records data about the type of card being read (whether it was a test for blood, semen,

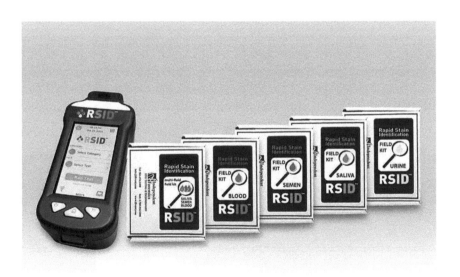

Figure 11.6   RSID test for semen.

Source: RSID and forensic handheld recorder, Google Search.

human saliva), date of the reading, and the result of the reader. The reader also creates an image of the card that it reads. The images are stored in the handheld reader and can be downloaded using a cable that connects the reader to a computer (**Figure 11.6**).

11.7 Presumptive Tests for Saliva

11.7.1 Phadebas Test

A chemical reagent called Phadebas is used to detect the enzymatic activity of the alpha-amylase enzyme, which is found in saliva. However, this enzyme is also found in other organisms as well. Alpha-amylases from bacteria, fungi, or chimps are very similar in structure and function to that of human alpha-amylase. Also, in humans, there are four variants of alpha-amylase, two of which are found in saliva, and the other two are secreted by the pancreas. And this is the reason this test is only presumptive, because it will give a positive result if the alpha-amylase enzyme is from any other organism present.

11.8 RSID Confirmatory Test for Human Saliva

The RSID test for human saliva detects the alpha-amylase molecule itself and, specifically, the alpha-amylase from human saliva. Performing both tests is considered confirmatory.

Of note, this RSID test has produced positive reactions in samples containing alpha-amylases from mammals such as gorillas and rats. This RSID also gave positive reactions from other bodily fluids, such as semen, blood, vaginal discharge, sweat, breast milk, and human feces. Due to this issue, as of January 2017, the NCSCL no longer conducts confirmatory testing for saliva.

11.9 ABO Typing

In the preceding section we have mentioned antigen and antibodies. ABO blood typing is essentially based on an antigen–antibody reaction that is carried out between red blood cells (RBCs), which serve as antigen (Ag), and serum (cell-free fluid) that contain antibodies (Abs). In scientific terms, Ags are substances that the body does not recognize as belonging to "self," and it mounts an immune response to make Abs to neutralize them. It is like getting a COVID-19 vaccine which contains Ags of the spike COVID-19 proteins and the recipients mount immune response to the spike proteins of the virus and develop antibodies (Abs).

When RBC antigens meet antibodies against them, they form clumps of RBCs. This process is called **agglutination**. More than 50 antigens have been identified on erythrocyte membranes, but the most significant, in terms of their potential harm to patients, are classified in two groups: the ABO blood group and the Rh blood group.

11.10 The ABO Blood Types

Although the **ABO blood group** name consists of three letters, ABO blood typing designates the presence or absence of just two antigens, antigen **A** on the membrane surface of RBCs type A, and antigen **B** on the membrane surface of RBCs B. Individuals whose RBCs have **A** antigens on their membrane surfaces are designated blood type A, and those whose RBCs have B antigens are blood type B. Some individuals can also have both **A** and **B** antigens on their RBCs; they are blood type **AB**. Individuals whose RBCs lack both **A** and **B** antigens are designated blood type **O**. ABO blood types are genetically determined, and therefore, ABO typing plays a very small part in forensics.

Here are a very interesting and surprising facts about ABO typing: individuals with type A blood have antibodies to the B antigen circulating in their blood plasma or serum. These antibodies, referred to as anti-B antibodies, will cause agglutination of RBCs which express B antigens on their membrane surface. Similarly, an individual with type B blood has anti-A antibodies.

Individuals with type AB blood, which have both anti-gens, do not have antibodies to either A or B antigens in their serum or plasma. Individuals with type O blood lack antigens A and B on their RBC surface but have both anti-A and anti-B antibodies circulating in their blood plasma.

11.11 Rh Blood Groups

The **Rh blood group** is classified according to the presence or absence of a second RBC antigen identified as Rh. Rh antigen was first discovered in the rhesus monkey macaque, hence the Rh antigen. There are several Rh antigens, but we generally type only one, designated **D** antigen. Those who have the Rh D antigen present on their RBCs are labeled as Rh positive (Rh+), and those who lack it are labeled as Rh negative (Rh-). The Rh group is separate from the ABO group, so any individual, regardless of their ABO blood type, may have or lack the Rh antigen. When identifying a patient's blood type, the Rh group is designated by adding the word *positive* or *negative* to the ABO type. For example, A positive (A+) means ABO group A blood with the Rh antigen present, and AB negative (AB-) means ABO group AB blood without the Rh antigen (**Figure 11.7**).

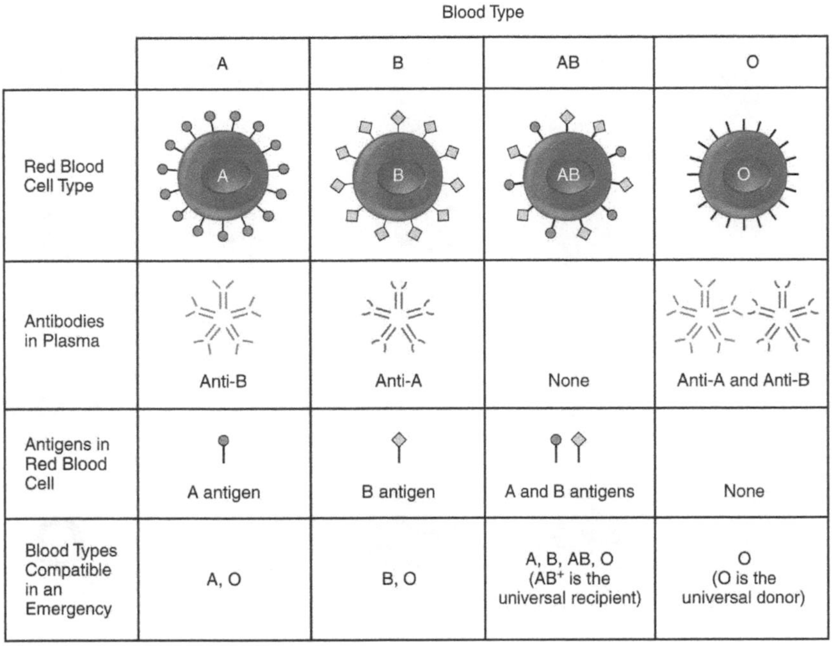

Blood Type

	A	B	AB	O
Red Blood Cell Type	A	B	AB	O
Antibodies in Plasma	Anti-B	Anti-A	None	Anti-A and Anti-B
Antigens in Red Blood Cell	A antigen	B antigen	A and B antigens	None
Blood Types Compatible in an Emergency	A, O	B, O	A, B, AB, O (AB⁺ is the universal recipient)	O (O is the universal donor)

Figure 11.7 The preceding figure illustrates the presence or absence of A and B antigens and the presence of antibodies in their plasma (serum). The bottom two rows show Rh antigen.

11.12 Frequency of ABO Types

In the United States, group O is the most common blood type, O+ accounting for 37.4% and O- are 6.6%; 42% of people are A type, while 35.7% A+, and 6.3% A-. About 10% are B type, with 8.5% B+, and 1.5% B-. The AB blood type is present in 4%, with 3.4% AB+ and 0.6% AB- (**Figure 11.8**).

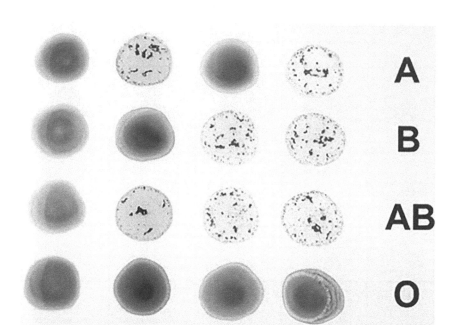

Figure 11.8 The previous figures illustrate the agglutination reaction between RBCs and their respective antibodies.

Chapter 12

How to Extract DNA from a Biological Sample at a Crime Scene

12.1 Genomic DNA Extraction: Principle, Steps, and Functions of Reagents

DNA extraction from a forensic sample is a process of purifying DNA. The sample can be tissue (body parts, bone, teeth, or any other piece of a human body), blood, saliva from a swab, cigarette butt, vaginal secretion (in the case of sexual assault), hair follicles, seminal fluid, spinal fluid, amniotic fluid, etc. The notion of extracting DNA from any cell is very simple. First, one must disrupt the cell membrane of the cells to expose DNA and then separate it from the rest of the cellular materials. The following will concentrate on isolating the

DOI: 10.1201/9781003182498-13

Figure 12.1   DNA extraction.

genomic/chromosomal DNA found inside the nucleus of eukaryotic cells, not the mitochondrial DNA from the hair or other cells (**Figure 12.1**).

12.2 Principle of DNA Extraction

The basic idea is to break down the cell membrane either by using physical techniques or by chemical methods to get a fluid containing all the cell components, including DNA. The process is achieved simply by lysing the cells, which results in cell lysate or cell fluid. During cell lysis, various chemicals and reagents are used to break down different components of cell. Therefore, RNA is broken down by using RNase, lipids by detergents and surfactants, and proteins by protease. After that, the lysate is treated with a concentrated salt solution to create a clump of all the broken components, which brings DNA into the solution. After this process, the whole mixture is centrifuged, which separates the clumped debris from DNA. After centrifugation, the debris settles down at the bottom of a conical tube and the DNA is left in the solution part. This solution is transferred into a fresh tube, and DNA is precipitated by adding ice-cold alcohol. Salt is added to increase the ionic strength, which speeds up the precipitation process. A pellet of DNA is obtained upon centrifugation of this solution. The supernatant is discarded except for the DNA pellet, which remains stuck to the walls of the conical plastic tube. The DNA pellet is resuspended either in slightly alkaline buffer or ultrapure water for subsequent use generally for PCR. But before that, DNA quantification is carried out to determine how much has been obtained. Typically, we only need about 1 ng of DNA to carry out the complete GlobeFiler™ profile. A hair follicle can give 1 ng of DNA, whereas saliva, vaginal secretion, and other specimens can give between 1 and 1,000 ng or more, depending on the amount of the sample submitted to a crime laboratory.

12.3 Cell Lysis: Lysis Buffer, Detergents, Reducing and Chelating Agents

There are different buffers available for different kinds of tissues and cells. For mammalian cells, SDS lysis buffer is used for cell membrane disruption. SDS is an anionic detergent, and it disturbs membrane proteins and lipid bilayers, resulting in the disruption of the membranes of all intracellular organelles (i.e., lysosomes, mitochondria, nucleus, endoplasmic reticulum, etc.). 2-mercaptoethanol (2-ME) is the most common reducing agent, and it breaks all the disulfide bonds between different polypeptides of a protein, leading to their denaturation. The cellular cytoplasm may contain DNase that can degrade the DNA. Therefore, we use EDTA as a chelating agent, and it binds to Mg^{2+} ions, which act as cofactors for DNase, and quenching Mg^{2+} ions deactivate DNase activity and save the DNA from degradation.

12.4 Tris Buffer

DNA is pH-sensitive and can be degraded by pH change. Tris acts as a pH stabilizer during the cell lysis process. It maintains the pH at 8. The buffer contains cations Na^+ or K^+, which bind to negative phosphate groups of

DNAs and make it more stable in an aqueous solution. In the absence of Na^+ or K^+, DNA molecules repel each other, and the double helix of DNA molecules cannot be formed. The DNA must be separated from the rest of the cell components. There are three major types of DNA separation from cell debris.

12.4.1 Ethanol Precipitation

Ice-cold ethanol is added to the solution containing DNA and cell debris. Proteins get dissolved in ethanol. Upon centrifugation, DNA is obtained in the form of a pellet on the bottom of the conical-shaped plastic tube. The supernatant containing cell debris is discarded, and in the following step, the DNA pellet is washed with ethanol to remove any salts or impurities. The tube is recentrifuged; the supernatant is discarded. The pellet is air-dried, and then the pellet is resuspended in pure water or appropriate buffer for storage. Instead of ethanol, isopropanol can also be used to precipitate DNA.

12.4.2 Phenol–Chloroform Extraction

A mixture of phenol, chloroform, and isoamyl alcohol (25:24:1) is added to the solution. Upon centrifugation, two distinct phases are obtained with a white interface between them.

1. The aqueous phase contains the DNA.
2. The organic phase (phenol–chloroform) contains the broken-down proteins, lipids, and other cell debris.
3. The interface contains white fragments of lysed proteins.

12.4.3 Minicolumn Purification

Minicolumn purification is a solid-phase DNA extraction method in which first the DNA is bound to a silica column by spinning, then DNA is eluted by using an appropriate elution buffer. After successful isolation from broken proteins (enzymes, histone proteins, etc.), lipids, salts, polysaccharides, and phenols, DNA is resuspended in Tris-EDTA (TE) buffer or pure water for further testing.

12.5 DNA Quantification

After DNA extraction from a biological sample, it is essential to quantify the DNA before using it for STR or other analyses. There are a variety of methods available for the quantification of DNA, including absorbance, agarose gel electrophoresis, and fluorescent DNA-binding dyes. The most used method is a measurement of the absorbance of the DNA sample by an ultraviolet (UV) spectrophotometer. DNA has a maximal absorbance near 260 nm; therefore, UV light of this wavelength is passed through the samples.

The higher the absorbance, the greater the concentrations of DNA in the extracted samples. This method can also measure the quality of the DNA. The instruments are highly sophisticated and measure the absorbance at 280 nm to determine the protein contamination at 280 nm. The A260/A280 ratio analyzes the purity of the DNA samples, and values of 1.8 or higher are related to the degree of purity of DNA samples. A disadvantage of this method is that single-stranded DNA (ssDNA) and RNA also absorb UV light at 260 nm, and you can therefore overestimate the amount of double-stranded DNA (dsDNA) concentration. A wide range of UV spectrophotometers is currently being used, varying from traditional instruments that quantify DNA in plates or cuvettes, which require larger volumes of samples to "NanoDrop" (ThermoScientific) that are designed to quantify DNA from microvolumes of the sample. The NanoDrop spectrophotometer uses very small volumes, that is, 1 microliter (mL), and nowadays is the most used method worldwide.

There are several other methods besides direct UV measurement. One of these is a dye method. For example, PicoGreen® is a fluorescent stain that selectively binds to dsDNA. It has an excitation maximum of 480 nm, and the emission of fluorescence is measured at 520 nm. It has the advantage that the reported DNA concentration is an accurate estimation of the quantity of dsDNA. However, this method does not provide an estimate of DNA purity, that is, the level of protein contamination present.

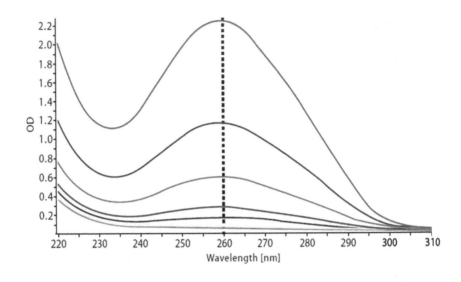

Figure 12.2   UV absorbance spectrum for different DNA concentrations at 260 nm wavelength.

In addition to that described previously, there are numerous other automatic instruments that a forensic laboratory uses. There are highly advanced kits that we use in the laboratory to extract DNA from minuscule amounts of biological samples. These kits are very simple to use and are very effective in DNA extraction (**Figure 12.2**).

Chapter 13

A Simple Guide for Understanding the Value of DNA Results

A Short Discussion on Probability and Statistical Calculations

13.1 Introduction

Once the DNA STR profiles are completed, the results of the analyses are presented in court. In addition to the DNA evidence, numerous other factors are involved in the legal course of proceedings. Therefore, eyewitness accounts, careful evidence collection, and chain of events, starting from the identification of the evidence and the authenticity of its collection until the final DNA results are resolved, involve tedious and meticulous medicolegal consideration. Statistical evidence plays a crucial role in

DOI: 10.1201/9781003182498-14

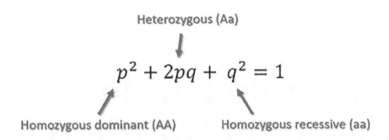

Figure 13.1   Hardy–Weinberg equation.

criminal investigations, prosecution, and court trials, not least in relation to forensic scientific evidence (including DNA) produced by expert witnesses. It is essential that everybody involved in criminal case resolution really understands the role that probability and statistics play in a court of law. Errors relating to statistical information and probabilities can result in serious miscarriages of justice (**Figure 13.1**).

In the United States, our court system is based on a sort of randomly selected jury, and very few of the individuals selected are equipped with an understanding of statistics or probabilities and how central this aspect of the court process is in the conviction or exoneration of an accused. Therefore, it is the responsibility of law enforcement personnel, judges, lawyers, and educated citizens to ensure that statistical evidence and probabilities are presented to the members of the jury in as clear and comprehensible a fashion as possible. It is sad to state that presently, even forensic scientists and expert witnesses, whose evidence is typically the immediate source of statistics and probabilities presented in court, may generally lack a clear understanding of statistics and probability and other relevant terminology, concepts, and methods. The main reason for this failure is that the STR statistical matches and subsequent statistical calculations are generated by computer-based algorithms. For example, if certain numbers of STRs match, the computer calculates the degree of probability automatically. These results are then presented to the judges, lawyers, and members of the jury. The expert forensic scientist who serves as the prime witness may not have a clear understanding of statistics and probabilities. Therefore, it is necessary that an expert witness satisfy the threshold legal test of competency before being allowed to testify or submit an expert report in legal proceedings. Unfortunately, most judges, lawyers, and public defenders have no idea of statistics, probability, or other simple subjects like this. Many lawyers, judges, and the public may have a fear of anything connected with probability, statistics, or mathematics, but irrational fears are no excuse for ignorance in matters of great importance. Their mistakes can ruin someone's life and can have a detrimental effect on their loved ones. These are busy folks and do not make time to learn one of the most essential missing gaps in their professional life. Or worse, others may be just plain ignorant of the lack of knowledge they must have to be fair to their clients, or arrogant enough to believe that they understand everything already. We believe that all the players involved in any criminal or civil case where forensic DNA results are presented should have enough knowledge of statistics and probability. The lawyers involved in forensic-related cases need to understand enough to be able to question the use of statistics or probabilities and to probe the strengths and weaknesses in the evidence presented to the court; judges need to understand enough to direct jurors clearly and effectively on the statistical or probabilistic aspects of the case. The expert witness must have enough knowledge to satisfy themselves that the content and quality of their evidence are appropriate to their professional status.

13.1.1 The Statistical Bases of DNA Evidence

One of the most distinctive features of DNA profiling, as compared with other branches of forensic science and forensic medicine, is that DNA evidence must be presented in statistical and probabilistic terms, and therefore, it is unequivocally probabilistic. There are numerous circumstances where DNA profiles would be useless or not permitted even were there a high statistical probability. Also, evidence cannot be presented in a criminal court unless it is relevant to a fact in issue. Without showing relevance, DNA evidence would not be admissible.

DNA evidence is always presented as proof of identity. Therefore, it is shown to identify an offender, a victim, or some other individual or individuals relevant to the case.

The identity of the culprit is sometimes the central issue in the court case, especially where the accused claims a mistaken identity or presents an alibi. DNA profiles are very useful when there is no clear eyewitness to a homicide, burglary, or rape. There will be cases where the DNA evidence will evaporate even with the highest significant statistical probability. For example, if one accused of rape admitted sexual intercourse and proclaimed that the complainant consented, DNA evidence will not provide much assistance to the prosecution in proving a charge of rape. Likewise, if the accused previously had legitimate access to a property, DNA collected from inside the property will not be evidence of theft, unless the DNA was recovered from a place around a broken window used to gain unlawful entry into the property. Again, DNA evidence does not rebut a claim of self-defense to a charge of assault, unless there is other evidence that proves inconsistencies in the accused's account.

It should also be clear that strong evidence forwarded by the prosecutor or the defense with a high probability of STR match is not always without question or doubt. There are numerous factors that play an important role in DNA analysis and its presentation to the courts, including the following: (1) genetic material from which a DNA profile could be generated remained at the crime scene, without irremediable degradation or contamination; (2) the physical sample was collected properly at the crime scene (or from the suspect and victim; (3) the sample was successfully transported to the laboratory without interference or contamination or tampering; (4) at the laboratory, the sample was analyzed using appropriately calibrated and properly functioning instruments and materials, with appropriate scientific protocols; (5) the results of the tests were accurately observed and recorded by a qualified forensic scientist; and (6) at no stage throughout the DNA procedures did the sample become contaminated with other genetic material, wrongly labeled, switched with other samples, etc.

Each stage of DNA handling must be documented and presented to the court. In the United States, the CSIs use temper-proof evidence collection packages. The forensic laboratory keeps a strict record of chain of custody, and they are periodically tested by the FBI with blinded DNA samples kept free from environmental contaminations. In summary, at the time of the DNA profile presentation, there should not be any doubt that the forensic DNA profile was generated from the same DNA samples collected at the crime scene, that it was not mislabeled and was free from contamination.

Most importantly, anytime a DNA sample is collected illegally due to serious police prejudice, or biased police investigation, such as with racial motives, it will be inadmissible as evidence. In the United States, extreme, biased police practices are very common, and a good defense attorney can make a case to exclude any evidence collected illegally or with prejudice. It should also be noted that eyewitness testimonies are only 3% accurate; many times, prosecutors and the police create eyewitnesses by reducing the sentences of individuals already incarcerated. A good defense attorney can also investigate this kind of unjust practice.

13.2 DNA Profiles as Evidence in Criminal Proceedings

In legal terms, biological evidence collected or recovered from a crime scene (i.e., blood, semen, saliva, etc.) is known as *"questioned"* samples, whereas the samples collected from a known person are known as *"reference"* samples. The reference samples can be from more than one individual—that is, from multiple suspects. Therefore, DNA profiles generated from a crime scene are known as "*questioned profiles,*" and those generated from a suspect will be a "*reference profile.*"

Reference samples are collected from a suspect under controlled conditions, often at a police station. Generally, it is a buccal swab, and it must contain a sufficient amount of high-quality DNA to generate an excellent and clear "reference profile."

On the other hand, a DNA profile from a "questioned" DNA can vary significantly, depending on the quality of the DNA. Usually, exposure of crime evidence to wet or hot conditions degrade the DNA, whereas in cold and dry conditions, DNA is relatively well-preserved. The evidence mixed with chemicals, such as dyes used in colored clothes, can interfere with DNA amplification, as can the age of evidence—how long ago the biological evidence was left on an object. The presence of mixed DNA—more than one person's DNA—can be problematic. For example, from a vaginal sample, there will be the victim's as well as the alleged rapist's DNA, but the sample may also contain DNA from the victim's husband or boyfriend.

13.3 The Value of Statistics and Probability in Forensic Profiles in the Court of Law

In a court of law, when DNA evidence is presented, it is NEVER absolute proof. There is no such thing as absolute, complete, and nonrevisable *certainty* in a court of law. Solid DNA evidence is presented in probability terms and as "beyond *reasonable*," not "beyond *all* doubt," or beyond "*every conceivable doubt.*" In simple terms, evidence is relevant when it affects the probability that a fact in an issue is true or false. Convicting evidence increases the **probability** that the accused is guilty. Conversely, evidence can also increase the probability that the accused is innocent. Vitally, evidence that neither increases nor decreases the probability that the accused is guilty is irrelevant and inadmissible.

13.4 The Value of DNA Evidence in Terms of Statistics and Probability

We have already covered the fact that each human being has a unique DNA profile based on STR except for monozygotic (maternal) identical twins. This "uniqueness" provides a powerful tool to the law in that each of us has a "UNIQUE" DNA profile.

Human DNA is shared by 100% of humans, and consequently, finding that a questioned sample contains human DNA does not discriminate between a particular suspect and any other human being on the planet. However, we know that the STR values for human DNA vary greatly between individuals, and it is this variation we calculate to create a probability number. The probable value of a DNA match is then open to probabilistic calculation, drawing on statistical calculations, including estimates of the relative frequencies of genotypes within ethnic subpopulations. For example, a profile of a "questioned STR" (i.e., *D3S1358*) matched perfectly with a "reference STR." Now, we will feed this information into the allelic frequency of *D1S1656* in a particular population.

To calculate the probability of each of the matched STR alleles, we must know the population distribution of alleles at each locus in question.

If the genotype of the relevant evidence sample is:

- Different from the genotype of the reference sample for a suspect, then the suspect is *excluded* as the donor of the biological evidence that was tested. An exclusion is independent of the frequency of the two genotypes in the population.
- The same as the genotype of the reference sample for a suspect, then the suspect is *included* as a possible source of the evidence sample.

The probability that another, unrelated individual would also match the evidence sample is estimated by the frequency of that genotype in the relevant populations.

All the DNA filers (i.e., GlobalFiler™, *AmpFℓSTR*™, *NGM SElect*™ kits) contain extensive loci. The database contains the frequencies of each of the STRs for various ethnic groups: African American, Asian, Caucasian, and Hispanic. In addition to the alleles recorded in databases, other alleles are published or reported to laboratories (see the STRBase at www.cstl.nist.gov/div831/strbase). Therefore, each of the matched STRs from a "questioned profile" and the "reference profile" is fed into the database, and the frequencies of each of the "matched" alleles are calculated by the software.

The **Table 13.1** that follows shows the frequencies for STR repeat numbers in different US populations. Once the frequencies of each of the alleles are determined, we can easily calculate the probability of a particular combination of STRs by multiplying the frequencies.

Table 13.1 Frequencies of Three STRs in Caucasian, African American, and Hispanic Populations

CODIS STR	No. of Repeats	Frequency (Caucasian)	Frequency (African American)	Frequency (Hispanic)
D3S1358	14	0.094	0.089	0.079
	15	0.111	0.186	0.293
	16	0.200	0.248	0.286
	17	0.281	0.242	0.204
	18	0.200	0.155	0.125
TH01	5	0.002	0.004	0
	6	0.232	0.124	0.214
	7	0.190	0.421	0.096
	8	0.084	0.194	0.096
	9	0.114	0.151	0.150
D18S51	10	0.008	0.006	0.004
	11	0.017	0.002	0.011
	12	0.127	0.078	0.118
	13	0.132	0.053	0.111
	14	0.137	0.072	0.139

Source: Data from Butler et al. (2003). *J Forensic Sci.* 48(4): 1–4.

The examples that follow show how we calculate probabilities and then multiply each of the exact matches to create a statistical table.

What is the probability of a Caucasian American having a 16, 17 combination for *D3S1358*?

0.200 × 0.281 = 0.0562
5.62% chance that a Caucasian American has a 16,17 combination for D3S1358.

What is the probability of an African American having a 16, 17 combination for *D3S1358*?

0.248 × 0.242 = 0.06
6% chance that an African American will have a 16,17 combination for D3S1358.

What is the probability of a Caucasian American having a 16, 17 combination for *D3S1358*, and a 5, 9 combination for *TH01*?

0.200 × 0281 = 0.0562
0.002 × 0.114 = 0.000228
0.0562 × 0.000228 = .00001281
0.00128% chance that a Caucasian American will have a 16, 17 combination for *D3S1358* and a 5.9 combination for *TH01*.

What is the probability of a Caucasian American having a 16, 17 combination for *D3S1358*, a 5, 9 combination for *TH01*, and an 11, 14 combination for *D18S51*?

0.200 × 0.281 = 0.0562

0.002 × 0.114 = 0.000228

0.017 × 0.137 = 0.002329

0.0562 × 0.000228 × 0.002329 = 2.98 e-8

0.00000298% chance that a Caucasian American will have a 16, 17 combination for *D3S1358*, a 5, 9 combination for TH01, and an 11, 14 combination for *D18S51*.

The probabilities significantly decreased when the number of CODIS markers are used combined with the probability data.

A female eyewitness has identified a Hispanic male as having stolen her car. The eyewitness noted that the man who stole her car was bleeding profusely from a head wound. Her car was recovered, and male blood with a 16, 17 combination for *D3S1358*, a 5, 9 combination for *TH01*, and an 11, 14 combination for *D18S51* was found on the driver's seat and steering wheel. Does this finding call the eyewitness evidence into question?

13.5 Here Is the Explanation in Probability Terms

The female eyewitness was wrong; a Hispanic male did not steal her car. The probability is that a Hispanic male will have a 16, 17 combination for *D3S1358*, a 5, 9 combination for *TH01*, and an 11, 14 combination for *D18S51*.

$$0.286 \times 0.204 = 0.058$$
$$0 \times 0.150 = 0$$
$$0.011 \times 0.139 = 0.0015$$
$$\mathbf{0.058 \times 0 \times 0.0015 = 0}$$

There is a zero probability that a Hispanic man stole her car (Table 13.1).

13.6 Summary

In brief, when a particular DNA sample is analyzed by using STR markers, we count the number of repeats of each STRs (all 13 CODIS STRs). For example, in the profile that follows, the STR marker FES has one strand of DNA with 8 repeats of the code word, and the other with 11 (each is inherited, one from the mother's DNA and one from their father's), and so has genotype 8/11. But the STR marker D3S1358 has 15 repeats on both strands and so has the genotype of 15. Once we analyzed and when we line up each of the 13 STR and the numbers of the repeats STR, the probability of all 16 STR matching is statistically one in several trillions, and that is the real DNA match!

Remember that population data and frequencies of each of the STRs are calculated according to the racial group and that final calculations, including the probability and statistical information, are presented to the court.

Marker	Genotype
FGA	20/24
FES	8/11
TH01	7/9
VWA	15/18
D3S1358	15
TPOX	8/10
CSF1PO	11/12
D5S818	12
D13S317	11/13
D7S820	8/9
D16S539	12/13
D2S1338	24/25
D8S1179	12
D21S11	30/33.2
D18S51	14/22
D19S433	14/14.2

Figure 13.2   Example of STR frequency.

The probability of a particular multiple-locus genotype is calculated by multiplying the frequencies of each of the STR loci (alleles) and including a factor of 2 for each of the heterozygous locus. The profile frequency is referred to as the **random match probability.**

Once this is stated, remember, blood-related individuals can share a lot more STR than unrelated individuals (**Figure 13.2**).

13.7 Probability and Statistical Significance Among Blood Relatives

The probability value of DNA evidence is reduced if the perpetrator and the defendant are blood relatives. While members of all genetically linked populations share distant coancestry, blood relatives have many closer common ancestors. The more DNA they share, the smaller the probability value of DNA profiles in distinguishing between blood relatives.

When we calculate likelihood ratios for unrelated individuals, we estimate the probability of their having shared alleles purely by chance. This is the genotype probability based on allele counts. Since the alleles are inherited from both parents, who inherited theirs from their parents, and so on, they back up the line of genetic descent.

For example, siblings are more likely to share inherited alleles from their parents than are first cousins, as illustrated in the following:

In **Figure 13.3,** males are shown by squares, and females by circles. Each arrow represents one allele

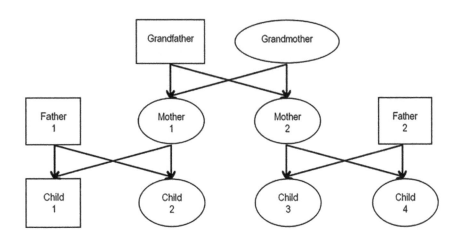

Figure 13.3   Family tree of siblings.

donated by a parent. You may recall from Chapter 2 that a parent is equally likely to pass on either of their two alleles to their offspring, and segregation and independent assortment plays a central role in inheritance. Consequently, the probability that a particular allele will be passed on to a particular child is halved in each successive generation.

The following figure illustrates the types of blood relationships most frequently encountered in forensic casework, with the percentage of pairs of individuals sharing none, one, or two inherited alleles at any locus. Identical twins and non-blood relatives are included for purposes of comparison (**Figure 13.4**).

FAMILY RELATIONSHIP	AVERAGE DNA SHARED %
Identical Twin	100%
Parent / Child	50%
Full Sibling	50%
Fraternal Twin	50%
Grandparent / Grandchild	25%
Aunt / Uncle / Niece / Nephew	25%
Half Sibling	25%
First Cousin	12.50%
Half Aunt / Half Uncle / Half Nephew / Half 1st Cousin	12.50%
Half-First Cousin	6.25%
1st Cousin Once Removed	6.25%
2nd Cousin	3.13%
2nd Cousin Once Removed	1.56%

Figure 13.4 Percentage of DNA shared by siblings and uncle, aunt, niece, and nephew.

At one end of the continuum, identical twins share both alleles by direct inheritance; at the other pole, people that are not blood-related have no directly inherited alleles in common. Siblings are more likely than first cousins to have the same alleles because they have inherited their alleles from their common parents. However, siblings are just as likely to have no alleles in common as to share both alleles (there is a 0.25 or 25% probability of either eventuality). By contrast, there is a 0.75 (or 75%) probability that first cousins would share no inherited alleles, and a 0.25 (or 25%) probability that they have one shared allele. These simple calculations demonstrate why genetic proximity reduces the size of the likelihood ratio and correspondingly decreases the probable value of a matching DNA profile.

Chapter 14

What Are Autosomal Dominant, Recessive, and X-Linked Dominant and Recessive Traits?

In this final chapter, our goal is to explain what autosomal dominant, autosomal recessive, and X-linked dominant and recessive genes are, how one inherits these genes, and what kind of results one can observe.

Just remember that genes act in pairs, one from the mother and one from the father. Gene pairs separate during meiosis, and sperms and eggs carry only one gene during the segregation and independent assortment process. When the sperm fertilizes the egg, the father's genes, located on his chromosomes, join the mother's genes, located on her chromosomes. Therefore, both contribute to the genetic makeup of the newborn.

14.1 Autosomal Dominant Inheritance

In the case of autosomal dominant inheritance, one gene is dominant over another gene, and the dominant form of the gene will be expressed in the newborn.

By analyzing a pedigree, one can determine genotypes, identify phenotypes, and predict how the traits are passed from parents to their children. This process allows one to figure out if a gene is autosomal dominant, recessive, X-linked dominant, or recessive.

It should be remembered that the presence of many affected children in a family is not always due to the dominant traits. If you use the Punnett square method, you will be able to easily determine that 50% of the children would be affected.

Examples of autosomal dominant diseases are Huntington's disease (involuntary movement, emotional disturbance, dementia), tuberous sclerosis complex, Marfan syndrome (long, thin extremities and fingers), neurofibromatosis (pigmented spots on the skin, skin tumors), polycystic kidney disease, and achondroplasia (dwarfism, large head, short extremities, short fingers and toes), just to mention a few (**Figure 14.1**).

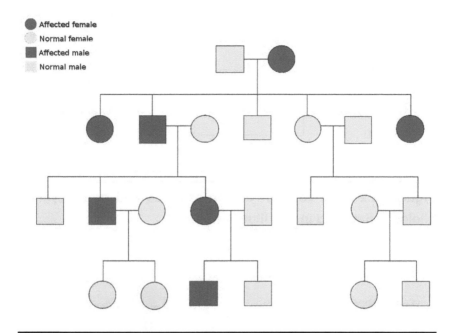

Figure 14.1   Autosomal dominant illustrated.

14.2 Autosomal Recessive Inheritance

An autosomal recessive trait requires a pair of genes, where both genes play an equal part; therefore, these are homozygous genes, as shown in the illustrations that follow. In the case of an autosomal recessive trait, 25% of the children would carry the gene. Nearly 2,000 trials have been related to single genes that are recessive, meaning, that their effects are masked by the normal gene. The most common example is sickle cell anemia, beta thalassemia (mild to severe anemia, stunted growth, bone deformation), albinism (lack of skin pigment, hair, and eyes, with significant visual problems), and cystic fibrosis (chronic lung and intestinal problems), just to mention a few (**Figures 14.2** and **14.3**).

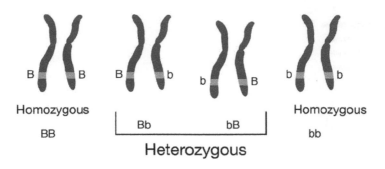

Figure 14.2   **The concept of homozygous and heterozygous alleles is depicted in this illustration.**

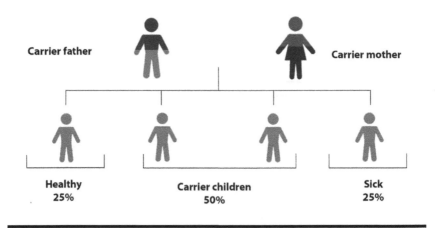

Figure 14.3   **The concept of autosomal recessive inheritance is illustrated.**

14.3 X-Linked Dominant Inheritance

In the case of x-linked inheritance, all daughters of a male who has the trait have the trait, whereas there is no male-to-male transmission of the gene. The traits follow the transmission of the X chromosome (**Figure 14.4**). The sons can have the trait only if their mother also has the trait. X-linked dominant, there is no question of a carrier, since the affected allele will be dominant (**Figure 14.5**).

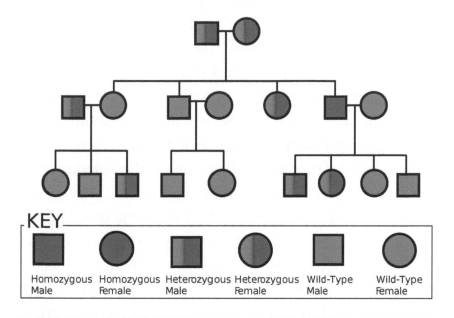

Figure 14.4   The concept of autosomal recessive inheritance is illustrated in the previous figure.

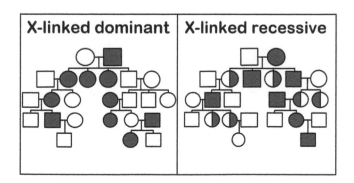

Figure 14.5   The following illustration shows both X-linked dominant versus X-linked recessive inheritance patterns.

14.4 X-Linked Recessive Inheritance

In the case of **X-linked recessive**, the carrier will always be the mother/female. Since she has two X chromosomes, one can compensate for the function of the other nonfunctional chromosome. The father/male can never be the carrier, because he has only one X chromosome. This trait is more common in males than females. If the mother has the trait, all her sons will have the trait. It has the same inheritance patterns as autosomal recessive for human females (not males). The sons of the female carrier would have a 50% chance of having the trait. Mothers of males who have the trait can be heterozygous or homozygous carriers; the latter will express the trait. Human disorders linked to X-linked recessive inheritance include hemophilia (bleeding tendency with joint involvement), fragile X syndrome (mental retardation, characteristic facies), and Lesch–Nyhan syndrome (cerebral palsy, self-mutilation), just to mention a few (**Figure 14.5**).

Bibliography

1. Steven D. Garber. Biology: A Self-Teaching Guide (Wiley Self Teaching Guides) Publisher Jossey-Bass; 3rd edition (August 21, 2020).
2. Bruce Alberts. Molecular Biology of the Cell. (July 2022). Publisher: W. W. Norton & Company; Seventh edition.
3. Elizabeth Cregan. All About Mitosis and Meiosis: Life Science (December 2007). Publisher: Teacher Created Materials; 1st edition.
4. Ann Felice Angeles. Adventures in the Cell: An Introduction to Mitosis and Meiosis (December 2020). Publisher: Independently published.
5. John Butler. Fundamentals of Forensic DNA Typing (First Edition Sept 2009) Publisher: Academic Press.
6. John Butler. Forensic DNA Typing: Biology and Technology Behind STR Markers (February 2001). Academic Press.
7. Bridget Hoes. Blood, Bullets, and Bones: The Story of Forensic Science from Sherlock Holmes to DNA. Publisher: Balzer + Bray; Reprint edition (February 20, 2018).
8. Mason Anders. DNA, Genes, and Chromosomes. Publisher: Capstone (August 2017).
9. Herman E. Wyandt, Golder N. Wilson, et al. Human Chromosome Variation: Heteromorphism, Polymorphism and Pathogenesis (April 2017). Springer; 2nd ed. 2017 edition.
10. Waleeb K. Heneen and Andrew S. Bajer. The Beauty of Chromosomes. Publisher: CreateSpace Independent Publishing Platform (November 10, 2017).

11. John Duffy. Chromosome 23 (July 2013). Publisher: CreateSpace Independent Publishing Platform.

12. Claudia Behrend, Javad Karimzad Hagh, et al.Human Chromosome Atlas: Introduction to diagnostics of structural aberrations (Jul 24, 2017). Publisher: Springer.

13. Jeffrey L. Shultz. Genetics: Includes the Central Dogma, Transmission Genetics and the Hardy-Weinberg Assumptions.

14. Lisa Spock, Tara Rodden Robinson. Genetics for Dummies (3rd Edition. January 2, 2020). Publisher: For Dummies.

15. Manfred Kayser. Forensic use of Y-Chromosome DNA: a general overview. Human Genetics. 2017: 136(5): 621–635. Publisher: Springer.

16. Fernanda M. Garcia, et al. Forensic applications of markers present on the X chromosome. *Genes (Basel)*. 2022 Sep 7;13(9):1597. doi: 10.3390/genes13091597.

17. Mohammed H. Albujja, et al. A review of studies examining the association between genetic biomarkers (short tandem repeats and single-nucleotide polymorphisms) and risk of prostate cancer: the need for valid predictive biomarkers. *Prostate Int*. 2020 Dec;8(4):135–145. doi: 10.1016/j.prnil.2019.11.003.

18. Iva Gomes, et al. Twenty years later: A comprehensive review of the x chromosome use in forensic genetics. *Front Genet*. 2020 Sep 17;11:926. doi: 10.3389/fgene.2020.00926.

19. Jeremy Watherston, et al. Current and emerging tools for the recovery of genetic information from postmortem samples: New directions for disaster victim identification. *Forensic Sci Int Genet*. 2018 Nov;37:270–282. doi: 10.1016/j.fsigen.2018.08.016.

Index

Page numbers in *italics* indicate a figure on the corresponding page.